THE OWNER-BUILT HOME

THE OWNER-BUILT HOME

by Ken Kern

Charles Scribner's Sons, New York

Copyright © 1972, 1975 Ken Kern

Library of Congress Cataloging in Publication Data

Kern, Ken, 1927–
   The owner-built home.

   Bibliography: p.
   Includes index.
   1. House construction—Amateurs' manuals.
I. Title.
TH4815.K47   690'.8   75-5653
ISBN 0-684-14218-X
ISBN 0-684-14223-6 (pbk.)

3 5 7 9 11 13 15 17 19   C/C   20 18 16 14 12 10 8 6 4 2
7 9 11 13 15 17 19   C/P   20 18 16 14 12 10 8 6

Printed in the United States of America

Lovingly dedicated to my wife, Barbara

Special thanks to design associates John Raabe and Bob Brooks for their help: John provided the cover design, and Bob did the architectural renderings.

CONTENTS

Preface   ix
Introduction   1
   1. BUILDING SITE   11
   2. BUILDING CLIMATOLOGY   18
   3. VENTILATION   25
   4. SUMMER COOLING   32
   5. LIGHT AND SHADE   42
   6. SPACE HEAT   52
   7. CENTRAL HEAT   63
   8. FIREPLACE HEAT   73
   9. LANDSCAPE DESIGN   88
  10. THE PLAN   98
  11. THE FREE-FORM HOUSE   106
  12. THE COURT-GARDEN HOUSE   116
  13. GROUP-LIVING SPACE   122
  14. INDIVIDUAL-LIVING SPACE   127
  15. COOKING AND DINING   131
  16. INSIDE YOUR HOME   136
  17. ADOBE BLOCK   141
  18. PRESSED BLOCK   150
  19. RAMMED EARTH   157

20. STONE MASONRY   165
21. MASONRY BLOCK AND BRICK   179
22. CONCRETE   192
23. PRECAST CONCRETE PANELS   201
24. WOOD   208
25. WOOD-FRAME STRUCTURE   219
26. POLE-FRAME STRUCTURES   228
27. COMPOSITE MATERIALS   236
28. PLASTICS   248
29. SALVAGE MATERIALS   254
30. TOOLS   261
31. THE FOUNDATION   267
32. FLOORS   277
33. WALLS   288
34. WOOD ROOFS   296
35. MASONRY ROOFS   301
36. STAIRS   310
37. PLUMBING   316
38. WIRING AND LIGHTING   328
39. LIGHT AND COLOR   334
40. DO-IT-YOURSELF PAINTING   342
CONCLUSION   349
ABOUT BOOKS   359
INDEX 368

PREFACE

One trouble with a preface is that it is written by the author after the book is finished, but it is seen by the reader before he reads the book. Therefore, when I extend hearty thanks to Wendell Thomas for editing each chapter, the reader can appreciate the task only after seeing the text. At the outset it may prove meaningless for the reader to know that inspiration and encouragement for writing my book came from Mildred Loomis, Education Director of The School of Living. Many of the chapters originally appeared in Mildred's publications—The Interpreter, A Way Out, and The Green Revolution. Some chapters also originally appeared in *The Mother Earth News*, and I am grateful to editor John Shuttleworth for permission to use my material for this single-volume edition. My special thanks go to my associate, Bob Brooks, for his help in preparing some of the perspective sketches.

The owner-builder concept states the idea that everyone can and should build his own low-cost house. This concept has been a subject foremost in my mind for over thirty years. I began extensive housing research in the late 1940s while attending Architectural College at the University of Oregon. Years of foreign travel followed my exploratory training period, but discussion with other builders and designers of minimum-cost housing throughout the world did the most to coalesce my own thoughts on the subject. Finally, it all came together in the fifties when I first started writing about the owner-built home for The School of Living journals. As with so many other fringe publications, barely alive in the cracks and margins of an affluent society, my work, too, was discovered by way of a generous review by Lloyd Kahn in *The Whole Earth Catalogue*.

The appearance of this edition coincides with the realization of a life-long dream for a building research center. Our new, nonprofit corporation, Owner-Builder Publications, was established to disseminate housing and food-production research. We also supply professional consultation and direct assistance in building design and homestead planning to individual owner-builders. All royalties from the present edition of my book are channeled into this corporation to underwrite this important service.

INTRODUCTION

The Owner-Built Home is intended to be a *how-to-think-it* book. Alternatives to the professionally executed, contractor-built home are presented in text and through sketches. None of the author's sketches are meant to be more than schematic representations of a thinking through process for one's house-building project.

The author has long had the compulsion to express feelings and thoughts in regard to the home-building industry and the wish to do something constructive for the people who suffer under it—both the construction worker and the home buyer alike. No critic as yet comprehends entirely why our houses are so poorly constructed, why they look so abominable, why they cost so much for construction and for maintenance, and why they are so uncomfortable. Some critics blame the building contractors personally; others feel that the fault lies with urban codes and building restrictions. Some believe that expensive housing is due to the high interest rates charged by the banks; others blame the trade unions for hampering efficient construction. Every writer on the subject seems to fondle some pet corrective measure. And every year some noted architect develops a sure-fire technical solution to the housing problem. Even more often

1

the building-material manufacturers come up with a new wonder, an improved wallboard or window or what-not, which can be installed with a ten-minute saving in labor.

Everyone in the building industry appears to be busily engaged in making "improvements" in his personal area of concern, but quality makes a steady decline. The end product is as inadequate, unsatisfying, and costly a house as ever. The architect spends more and more time at his drafting board, exhausting possibilities of new construction techniques and more economical arrangements; the contractor conscripts ever more specialized equipment for building efficiency; the banker resorts to undreamt-of schemes to make it possible for everyone to buy his new home—even if he lacks money to make the down payment; building-material manufacturers work overtime in their laboratories making "more and better things"—presumably for better living. With all this bustle one might well expect some major improvements in new-home construction. Whatever improvements have occurred are insignificant in comparison with the improvements that should be made. The causes of the world's housing problem still remain.

Tracing these causes to their sources has helped me to view the problem in perspective—comprehensively. This process has also suggested some workable alternatives as solutions to personal housing needs. Here they are in the form of seven axioms, listed in order of importance for the prospective owner-builder.

1. *When building your home, pay as you go.* A building loan is a type of legalized robbery. More than any other agency, banks have been successful in reducing would-be democratic man to a state of perpetual serfdom. The banks have supported and helped to determine social and political conventions and have amassed phenomenal fortunes through unearned increment. As "friends" of the homeowner they have made it possible for him to take immediate possession of his new home—and to pay for it monthly for 20 to 30 years. Most people who fall into this trap fail to realize that the accumulating interest on their 30-year mortgage comes to more than double the market value of their house! If one expects any success at all with keeping costs of his new home down to a reasonable price one must be entirely free from interest rates.

2. *Supply your own labor.* Building trades unions have received—and not unjustly so—a notorious reputation as wasters of speed and efficiency in building work. We all know that painters are restricted to a certain-size brush and that carpenters are limited to a certain-size hammer (upon threat of penalty from union officials). Apparently more width and more weight might conceivably speed up a project to the point where some union man would prove to be expendable.

The disinterest that the average journeyman has in his work, despite his high union-pay rate, is appalling. The lack of joy-in-work or of acceptance of responsibility among average workmen can be accounted for partly by the dehumanizing effect of the whole wage system. So long as the master-and-slave type of employer-employee relationship continues to exist in our society one can expect only the worst performance from his "hired help." Until the dawn of the new era approaches one would do well, from an economic as well as from a self-satisfying standpoint, to supply his own labor for his own home insofar as he can.

3. *Build according to your own best judgment.* At the apex of the poor-building hierarchy—and perhaps the greatest single impediment to good housing—is convention. Building convention takes two forms: first, there is convention that is socially instilled, commonly called "style," which can be altered through education. The second type of convention is more vicious and politically enforced. Building codes, zoning restrictions, and ordinances all fall into this class of impediments. In urban jurisdictions politically controlled convention calls the shots for practically every segment of the building industry. Ordinance approval or disapproval makes the difference between having a house or having none at all. Or it may make a difference of $1,000 (on the average) wasted because of stupid, antiquated building laws.

If we are to be at liberty to build our own home at less cost, we must necessarily be free from building-code jurisdiction. This means we must locate outside urban control—in the country or in small-township districts.

4. *Use native materials whenever possible.* Much of an architect's time is spent keeping abreast of the new, improved building materials, which manufacturers make each month. Many of the products are really worthwhile, but, more often than not, in cost they are entirely beyond the reach of the average home builder. Basic materials, like

common cement, have not appreciably advanced in price over the past dozen years, but some of the newer surfacing materials and interior fixtures have skyrocketed in price during this same period.

By not using these high-cost materials one, of course, avoids this problem. Emphasis should, instead, be placed on the use of readily available natural resources—materials that come directly from the site or from a convenient hauling distance. Rock, earth, concrete, timber, and all such materials have excellent structural and heat-regulating qualities when properly used.

5. *Design and plan your own home.* One ten-percenter with whom we can well afford to do without when building a low-cost home is the architect-designer-craftsman-supervisor. Experience in this aspect of home building has led me to conclude that *anyone* can and everyone *should* design his own home. There is only one possible drawback to this: the owner-builder must know what he wants in a home, and he must be familiar with the building site and the regional climatic conditions. Without close acquaintance with the site and the regional climate and without a clear understanding of the family's living needs the project is doomed to failure no matter who designs the house. An architect—even a good architect—cannot interpret a client's building needs better than the client himself. Anyway, most contemporary architects design houses for themselves, not for their clients. They work at satisfying some aesthetic whim and fail to really understand the character of the site or the personal requirements of their client.

6. *Use minimum but quality grade hand tools.* If the house design is kept simple and the work program is well organized an expensive outlay in specialized construction equipment can be saved. The building industry has been mechanized to absurd dimensions. And even with more and better power tools labor costs rise. At times, where labor savings occur, the difference is taken up by the depreciation and the maintenance of the equipment, which saved all the time in the first place. Whatever way one looks at it, a certain amount of work must go into building a home. If a prospective home owner is unprepared to accept the challenge of building his own home—and falls into the over-stocked power-tool trap—then he must be prepared to spend greater sums for a product which could very well prove to be inferior.

7. *Assume responsibility for your building construction.* The general contractor has become such a key functionary in practically every

building operation that one soon loses sight of the fact that he is a relative newcomer to the housing scene. Not many years ago the contractor's job was performed by a supervising carpenter, a so-called "master builder," who had control of the whole project. Once people realize how little is involved in implementing a set of building plans they will better appreciate the fact that the contractor is the most expendable element on any job.

Excessive profits are made by the general contractor for coordinating the work stages and for assuming the responsibility for satisfactory completion of the work at a specified cost. For this service he receives 10 percent of the total cost of your house. Besides, he receives an even greater percentage on all materials that go into the structure. The contractor is an expensive and a nonessential luxury for the low-income home builder.

Now that the ideal program for the owner-built home has been presented, steps should be retraced, and the sheer realities of the situation should be faced. Obviously, not all people can locate their home sites out of building-code jurisdiction, nor can many people expect to finance their homes from their weekly paycheck. Very few people have the native ability to design an inexpensive and an attractive home—one that truly fits their needs and site conditions. Even more rare is the person who can carry through all phases of building construction or who even has the necessary free time to devote to a house-building effort. How many people do you know who can take the raw-material resources and process them into building materials for wall, roof, and floor? One has only to observe current owner-built home flops to appreciate the fact that we are dealing with a disturbingly complex problem—a problem which demands a comprehensive solution.

It is unquestionably our drive toward specialization (stemming from a basic failure on the part of our whole educational system) which is primarily responsible for modern man's inability to provide directly for his own shelter needs. Despite this trend the owner-built home can be an economical as well as an aesthetic success. It has been so for centuries for millions of families, and it continues to be so today. Furthermore, the process of building one's home can become one of the most meaningful and most satisfying experiences in one's life— as, indeed, it should. Owing to the physical limitations of the owner-builder and to those impositions foisted upon him by society's restrictions and its general miseducation one can only expect to approach the completely self-tailored home. On one or more counts compromises are in order, but to the extent that the owner retains full control over his design and his work he is successfully participating in creative building.

Experience in the building design and in the construction fields in a number of countries has taught me one very important lesson: satisfactory progress with the low-cash-cost, owner-built home can come only after the occupant's adoption of an entirely new approach to materials, structure, finished appearance, and his own basic life style. Our existing ego-inflated, over-materialistic, and downright absurd housing forms are gross impediments to the sort of rational and economical building that is both possible and desirable. But to find intelligence in housing today one must go to the countries that, out of sheer necessity, are beginning to approach the housing problem at its roots.

In Asia, for example, 150 million families live in overcrowded and unsanitary quarters. Some countries, like India, are attacking this situation with energy and imagination. A series of Aided Self-Help programs are included in the Indian government's three-year Community Development plan. At the International Exhibition on Low Cost Housing held in New Delhi a few years ago, a complete model village was on display. Over 30,000 people visited this village each day. It proved to be the most successful low-cash-cost demonstration center in the world. None of the dwellings in this village cost over $1,000. Besides the wide variety of domestic buildings, the village contained a school, health clinic, co-op store, carpentry shop, and smithy. The village was laid out with proper regard to water supply, drainage, lighting, and street planning. This demonstration center also illustrated the wide variety of low cash-cost materials that are available: reeds, aluminum, gypsum, hessian, rammed earth, and concrete— each employed in new and more imaginative ways.

The new structural ideas, uses of materials, and methods of design which result from such an effort as that at the New Delhi exhibition mark a tremendous architectural advance—but the human advance behind the scenes is even greater. The best thinkers in their field have been on the job. Men like Kurt Billig, director of the Central Building

Research Institute (Roorkee), A. L. Glen, (Pretoria), and G. F. Middleton, Commonwealth Experimental Building Station (North Ryde, New South Wales, Australia) could command the highest fees from those most able to pay. Instead, they contribute their vast store of building knowledge and imagination to the greatest housing needs of our age. Architect Joseph Allen Stein, head of the Department of Architecture at Bengal Engineering College (India), summed up my sentiments when he made the following statement at the New Delhi exhibition:

> Centuries of privation, of social and economic inequality, have conditioned vast numbers of human beings to endure surroundings that can only be called subhuman. Today, architects, engineers, and planners are called upon to show that a pleasant, healthful, humane environment need no longer be the monopoly of a fortunate few.

It is a rarity of the first order when the dean of an architectural college takes it upon himself to build houses comprised of woven split bamboo placed between two layers of treated clay. These readily available materials were artfully used by Professor Stein in his creation of two demonstration low-cost homes. In his own words, the design

> was worked out so that under proper conditions of community organization, such buildings can be built by village families with their contributed labor, without dependence on extra-village materials—*on the basis of a program of guided self-help.* The skill required for this type of construction is readily acquired; a two-month's apprenticeship is usually considered time for a man to become a skilled bamboo worker.
>
> If properly used, bamboo-and-clay construction can be expected to last as long as many manufactured materials that are considered to make permanent industrial housing. Standard materials for urban construction, such as corrugated iron sheets, poorly burnt, inferior bricks, or unseasoned wood can hardly be expected to last twenty-five years under average urban conditions. Yet even in the extremely hot humid climate of West Bengal and Assam, there are many clay and bamboo structures of forty years age. When replacement or repair is required due either to accident or deterioration by age, the materials are readily at hand, and the householder himself can do the work. The roof is of such a design that

repairs can be made to any portion without affecting, or having to break up, the remaining part.

[The rural house] is constructed of only three materials: it utilizes wood for the roof framing; the remainder of the construction is of earth (clay) and bamboo. In villages where wood is not readily and cheaply at hand, bamboo can be substituted. The sole purchase from outside the village is creosote, or other preservative materials, desirable to prolong the life of the structure.

Some of the world's underprivileged countries maintain a caliber of low-cost-housing research which surpasses that of the far more wealthy countries, such as our own. More significant research material is coming out of the South African Research Institute, for instance, than from all the H.H.F.A., F.H.A., and F.P.H.A. agencies combined. A recent housing development in South Africa made use of such construction features as no-fines concrete (crushed stone and cement) for surface beds and single-thickness, brick, internal walls—plastered on both sides. Detailed investigations were made on every item of expense that went into the experimental house in this development.

In this hemisphere the most important low-cost, owner-built housing research is being done at places like the Inter-American Housing and Planning Center (Bogotá, Colombia) and The Minimal-Cost Housing Group, Department of Architecture, McGill University (Montreal, Canada). Some years ago the agency in Bogota built a demonstration soil-cement house at a cash cost of $375 (see illustration, p. 9). Designed for the cool climate prevalent on the Andean plains, the house has a living room, kitchen, two bedrooms, covered porch, storage room, shower, and laundry area, apart from an outside latrine. Roof members were constructed with eucalyptus tree limbs. Common clay tiles, used for the roof, were placed with a mud mixture on a frame of split bamboo. The floor was constructed of tamped earth, covered with a layer of weak cement and topped with soil-cement floor tiles. In Montreal, just last year, The Minimum-Cost Housing Group constructed a demonstration house for a materials cost of $1,901. It was built of interlocking blocks of sulfur (utilizing some of the vast supplies of wastes from the oil, copper, and zinc industries) and with a timber-wall system of cost-saving modular construction, using notched and grooved logs to form load-bearing walls. Asbestos sewer pipes

LOW-COST HOUSE. BOGOTA

were cut to make giant, self-supporting roof tiles, from which all available rain water could be drawn, and floors were constructed of sulfur tiles. (See illustration, page 10.)

In this book my approach to housing utilizes technical features similar to those of the low-cost-housing research mentioned above. In the following chapters an evolutional frame of reference is introduced. The kind of house proposed involves a process of full growth and development for its realization—not only from the first conception of design and plan to the final nail that is driven, but also from an internal growth and maturation on the part of the owner-builder. The end product is as different from the reactionary contractor-built, bank-sponsored, tract house as it is from the revolutionary architect-designed, owner-financed, suburban home. What distinguishes the proposed evolutionary form of owner-built home is its fitness for purpose and its pleasantness in use.

A positive philosophical outlook and way of life must necessarily precede the achievement of a quality owner-built home. This is to say that a truly satisfying home must *develop* from other and more subtle patterns. The mere technical problems of building a home are insignificant when compared to an understanding and an interpretation of one's innermost feelings and thoughts concerning his shelter needs. But, if these feelings and thoughts are not consistently related to and released in daily activity, or if they become life-negative in orientation, then one might just discount the prospect of creating a satisfying home. Thoreau said:

> What of architectural beauty I now see, I know has gradually grown from within outward, out of the necessities and character of the indweller, *who is the only builder*—out of some unconscious truthfullness, and nobleness, without ever a thought for the appearance, and whatever additional beauty of this kind is destined to be produced will be preceded by a like unconscious beauty of life.

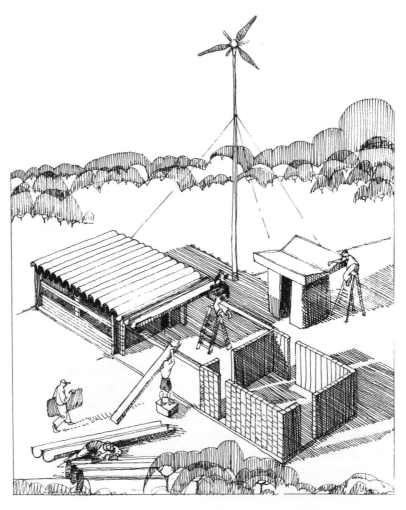

1 BUILDING SITE

Patrick Geddes should rightfully be considered the foremost as well as the first Western regional planner. One of his favorite expressions—and he had many of them—comes to mind as primary thought is begun on this book for the owner-builder project. Geddes was especially intent to make correct and proper first-beginnings, and he therefore counseled: *survey before plan before construction.*

An owner-builder could find no better advice to follow than this suggestion. The survey—of one's building intentions, of one's personal space needs, of one's building job, itself—must precede subsequent regard for building design, for plan layout, or for the actual work of construction. Survey, then, begins on the site. Property corners and boundary locations need to be established.* Following a clear demarcation of physical boundaries an owner-builder must locate and evaluate quite a number of physical features. In our program of assisting owner-builders with their building design we request a sketch of the building site showing property lines and site dimensions, ground slope direction and amount, latitude and solar orientation, public access and

*For simple survey instruction see *The Owner-Built Homestead,* chapter one.

preferred driveway location, distant and immediate view direction, summer breeze and winter-storm direction, existing or future neighboring buildings, utility location (water supply, electrical power, sewage), and existing trees, rocks, and other landscape features.

1.1 SITE ANALYSIS

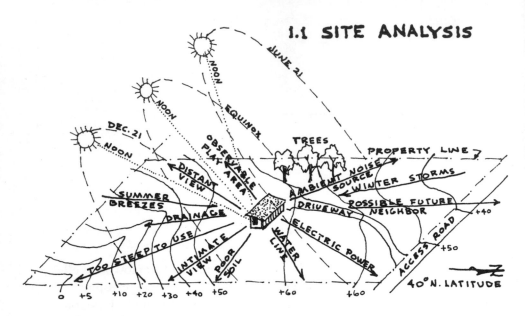

A schematic site-evaluation sketch along with a plan view of a possible site is shown above. These sketches illustrate some of the more important site conditions, which can and should play a dominant role influencing the design of the well-planned owner-built home. Influences of site on building design are little understood and little appreciated aspects of conventional building construction. Nevertheless, they are aspects which affect every person who uses the building. The realized design, in turn, affects the site, and these two features condition one's life and plans for years to come.

It seems entirely logical that every individually designed home should have more than the usual degree of site planning. Besides being expressive of its owner's life, a home should be at one with its site and the regional ethos. A man building his own home can afford to spend the time necessary to acquaint himself with the physiognomic-climatic site environment. The speculative or commercial builder usually fails to take enough time from his actual house-building program to know the character of the land upon which he is building. Results of this neglect are almost always unfortunate.

When the individual prospective home builder becomes acquainted with even a few of the specific site conditions found on his plot, he will begin to appreciate the fact that sites tend to vary as much as people do. No two sites are exactly the same, no two regions are the same; no two climates are the same. Hence, every building-design problem must be solved individually. One should add, of course, that no two persons are the same, nor do they have the same needs. We are dealing with three independent, though interrelated, components: people, site, and building. Both visually and actually, the building exists only in relationship to the site and to the surrounding landscape. In the same manner, the site exists in relation to the people—through the introduction of the house.

It is important to consider the house and the site as one indivisible whole. The house-planning and site-planning processes must happen together, with equal consideration to the design of every square foot of indoor-outdoor space. Lawns and workshops and gardens contain essences of their own, and it is as important to the total design concept that these be adequately expressed as it is that the essence of "living room" be expressed. It is something of a help to think of the house and site as a coordinated grouping of related indoor and outdoor rooms. In contemporary design work we are apt to concern ourselves with the psycho-physiological requirements of interior space and to exclude a consideration for the equally strong need that people have for a satisfying relationship with the outdoors. The control or lack of control of climate can be as important a design feature as the determination of the refinement of interior surface materials. One's relationship to view or to plants can be an extremely significant design feature.

The so-called contraspatial house grew out of this integrity-of-the-site concept. Other types, the binuclear house and the trifunctional house, have been gaining popularity in recent years. However, for every serious attempt to achieve integration of house to site you will find a thousand houses peppering the landscape and clearly demonstrating the builder's total disregard for even the most basic consideration of sun, wind, and view. In-between these extremes you will find scores of half-baked efforts which struggle to achieve some sem-

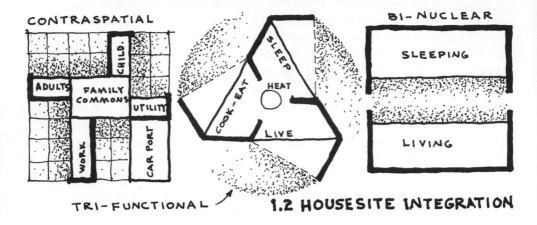

CONTRASPATIAL

CHILD.
ADULTS
FAMILY COMMONS
UTILITY
WORK
CAR PORT

TRI-FUNCTIONAL

SLEEP
COOK-EAT
HEAT
LIVE

BI-NUCLEAR

SLEEPING

LIVING

1.2 HOUSESITE INTEGRATION

blance of site-relationship. This writer is more critical of these latter abortive attempts than of the former disregard. The contractor-built tract home is, at least, an honest failure since it doesn't even try for integration. A few examples of the half-baked, modern efforts may suffice as forewarnings to the owner-builder in his approach to site planning.

The urge for a dramatic architectural effect usually impels the modern designer to place the structure in the most prominent position on the site. Or, for ease of construction and access, the house is located on the most level ground. The house can, however, be located on precipitous topography, often to great advantage. It is generally a mistake to build upon the most beautiful, most level section of the site. Once this area is covered with massive structures its original charm is destroyed.

The machine-for-living approach to house design and site planning is about as false to man's true living needs as is the art-for-art's-sake approach to his practical needs. In the former case, all important rooms in the house are oriented due south (in our northern hemisphere, of course)—irrespective of outlook or interior planning. The result sought by this selection is the achievement of maximum heat-gain in the winter and minimum heat-gain in the summer—at the expense, however, of having all rooms end up with the same lighting conditions inasmuch as all rooms will have south-oriented, glass fenestration.

Glass is one material which is very much misused by modern designers who respond to the bring-the-outdoors-in notion with floor-to-ceiling sheets of crystal. Paradoxically, just the opposite effect can be created, namely, claustrophobia, which results from one's urge to break the glass to get out! Obviously, the glass restricts an easy ingress and egress, though it succeeds by *suggesting* such movement.

The picture window is, of course, the epitome of the mistaken notion of bringing-the-outdoors-in now held by ding-bat contractors everywhere. Picture windows are to homes what show windows are to stores. They exemplify the marketplace mentality with its display of things. In essence, the picture window provides a vicarious experience: more people can sit in their armchairs to look at, not live with, nature.

One final example of the ways in which modern dwellings fail to integrate house and site involves view. When one is fortunate enough to have a site with a dramatic outlook, especially to the south or east (as mentioned in the machine-for-living approach cited above), the natural inclination is to orient all the major rooms toward that direction and to use glass in as much of this view-wall as is structurally feasible. A house so constructed speaks of arrogance and greedy self-importance. At best, the end result is unpleasant and distracting.

In this consideration of view we can learn much from Japanese builders. (Readers of this book will find frequent reference to Oriental architectural features. The author has long felt that the traditional Eastern forms have more to offer the modern-day owner-builder than most of our up-to-date source material.) A general practice among the Japanese is to place the house so that the same view is never seen from more than one vantage point—except in instances where the second view presents a contrasting element not seen by the first. A sequence of outlooks should be developed—from entry into the front yard and from entry into the house to a final view when stepping onto the outdoor terrace. The owner-builder needs to investigate the prospects for varieties of outlook and to employ some of the many devices for enhancing it. One interesting idea to develop is a contrast between the long view, such as that of a distant mountain range, and the short view, that of a patio garden. Again, it is unpleasant to view something perpendicularly through glass. The Japanese shy from the picturelike impression by offsetting the center-of-view interest and by creating hidden, around-the-corner vistas.

In his book, *Japanese House and Garden,* Dr. Jiro Harada gives the final word on view when he tells what Rikyu, a famed Japanese tea-master, did more than 360 years ago to give his garden deep, spiritual significance:

> When his new tea room and garden were completed at Sakai he invited a few of his friends to a tea ceremony for the house warming. Knowing the greatness of Rikyu, the guests naturally expected to find some ingenious design for his garden which would make the best use of the sea, the house being on the slope of a hill. But when they arrived they were amazed to find that a number of large evergreen trees had been planted on the side of the garden, evidently to obstruct the view of the sea. They were at a loss to understand the meaning of this. Later when the time came for the guests to enter the tea room, they proceeded one by one over the steppingstones in the garden to the stone water basin to rinse their mouths and wash their hands, a gesture of symbolic cleansings, physically and mentally, before entering the tea room. Then it was found that when a guest stooped to scoop out a dipperful of water from the waterbasin, only in that humble posture was he suddenly able to get a glimpse of the shimmering sea in the distance by way of an opening through the trees, thus making him realize the relationship between the dipperful of water in his hand and the great ocean beyond, and also enabling him to recognize his own position in the universe; he was thus brought into a correct relationship with the infinite.

The site plan and sketch illustrated on p. 12 cannot indicate what is perhaps one of the most important aspects of site planning, the site's physiognomy; that is, its essence or spirit—the original individuality of the site. If the owner-builder is fully aware of his particular site as it relates to the ethos of the regional landscape and to the character of the existing neighborhood, he will not go too far wrong in his site-planning practices. Much can be said about the human feeling toward the setting, especially with regard to one's immediate plot of ground—with regard to the microcosmos and the micro-climate of one's half-acre lot, let's say. Care and loving attention can have a major impact on this setting, and high quality site development can result where seemingly the only investment is one's imagination tempered by his full realization of the profound assets which lie within the site.

Ambient forces are apparently allowed to exert their full energy, unhampered—but, on the contrary, are developed—by personal redirection.

The best approach to site development lies somewhere between the masterful and the subservient levels. One should neither wreck the site nor fail to develop its character. Richard Neutra speaks of the consequences of disregard for the site's individuality:

> Try to understand the character and peculiarities of your site. Heighten and intensify its inner grain and fiber. You will pay dearly for any such offense, though you may never clearly note what wasting leak your happiness has sprung.

It is only after this feeling-for-the-site aspect has been realized that one should begin to plan the structural components of the site plan. Three general areas of space may be outlined: the public area, the private area, and the service area. Under each heading, list all the space requirements proper to it: a patio-garden living room, a game-play area for children, an outdoor work area (crafts, hobby projects, auto repair), outdoor storage facilities for garden tools and firewood, a trash area, plant structures (lathhouse, greenhouse, garden work center), a vegetable garden, fountain or swimming pool, perhaps someplace for animals.

As your desires and needs are listed and the space allotments for their satisfaction plotted on the site map the plan will thrive and begin to take form. Like a successful jig-saw puzzle each component will fall into its obvious, unmistakable position. You will know that a particular function must take place at a particular place on the plan, and that a certain amount of space must be allocated for another particular need. Soon the whole scheme will become immediately perceivable. It will be right, and you will be sure of its rightness. You will know when the time has arrived for the first stage of plant arrangement and of building design.

2 BUILDING CLIMATOLOGY

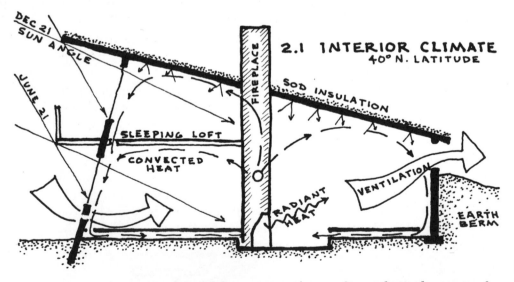

DEC 21 SUN ANGLE

JUNE 21

FIREPLACE

2.1 INTERIOR CLIMATE
40° N. LATITUDE

SOD INSULATION

SLEEPING LOFT

CONVECTED HEAT

RADIANT HEAT

VENTILATION

EARTH BERM

The cross-section sketch above represents a climatological survey of a hypothetical building interior. Analytical procedures for determining sun angle and air-movement direction are not unlike the survey of site conditions discussed in the previous chapter. We've merely taken the survey analysis indoors. References to Figure 2.1 will be made throughout this book: this sketch symbolizes the very best owner-builder thinking about climate control, simplicity of construction and functional design.

Building climatology is best introduced in this chapter by describing Harold Clark's environmental study of the subject done a number of years ago. This Columbia University professor visited forty countries during his inquiry. Upon his return, he told university students that he found few instances of private dwellings currently designed to suit the environmental climate. He deplored the fact that practically all modern dwellings throughout the world are patterned after the boxlike European houses, which fit the cold European climate. If Professor Clark could have conducted his environmental study a few centuries ago, however, it is certain that his concluding observations would have been more favorable since he would have, then, found that the indigenous and often primitive architectural forms of that time had become suited to local climate through a long process of trial and error.

Architecture these days ignores environment: witness the growth of the world's cities, which violate natural principles of summer cooling, for instance. Contrast the cool, shady woodland found in nature with the exposed acres of urban pavement, concrete buildings, and reflecting rooftops. Compatibility of the building with its environment is currently neglected as modern designers devote a disproportionate amount of attention to appearance and fashion—which, of course, boost the sale value of the package.

Owing to the extensive use made of climatic averages in describing regional climate conditions, there is a widespread tendency to regard climate as uniform in each latitude and in each season. Nothing, however, could be farther from the truth when dealing with actual climate and, especially, when dealing with building design. Human reactions to temperature depend upon the ability of the body to lose heat to one's surroundings by *convection* into the air, by *radiation* to the surrounding surfaces and by *evaporation of moisture* from the skin. Body reactions, therefore, depend not only upon the temperature of the air but also upon its humidity and its rate of movement—as well as upon the mean (or average) radiant temperature of the surrounding surfaces. It is utter nonsense to talk of a "72°F design temperature." A dry bulb of 72°F temperature at 90 percent relative humidity with a ten-foot-per-minute air movement will convey the same effective temperature as a 100°F bulb at a 10 percent humidity and a one-hundred-foot-per-minute air movement. In both instances the combination of meteorological factors will produce an *effective temperature* of 80°F in a room where the walls, floor, and ceiling are at the same temperature as the air. When the surrounding surfaces are *not* at air temperature, an altogether different temperature index is employed to measure the actual meteorological conditions. This "adjusted" index is called the Corrected Effective Temperature (C.E.T.).

The three basic climate relationships should, accordingly, be kept in mind. This will prove most helpful in cooling or in heating the owner-built home. They are:

1. Temperature is related to effective humidity. As temperature rises relative humidity drops. When high temperatures combine with high humidity the body has difficulty perspiring, and it experiences acute discomfort.

2. Air temperature is related to average radiant (or surface) temperature. In order to keep the body in an optimum-comfort zone with low-air-temperature conditions the radiant temperature must be kept high. In summer, when the air temperature is high, a low radiant temperature is required.

3. Air movement is related to both temperature and humidity. Up to a certain point high temperatures can be counteracted by air movement.

After discovering a way to incorporate the foregoing climatological principles into a single index, the American Society of Heating and Ventilating Engineers produced the Effective Temperature Scale. Even the Effective Temperature Scale has its limitations in terms of indicating actual body comfort. For instance, cold concrete floors in a room of otherwise comfortable effective temperature will produce discomfort by constriction of the blood vessels of the feet. If the feet lose heat rapidly from contact with a cold floor a person will experience discomfort, even though the "official" effective temperature is within the comfort zone. Conversely, it is possible to feel warm in a relatively cool room, if seated with feet out-stretched in front of an open fire. A high ceiling temperature, also in a comfortable effective temperature range, will produce uncomfortable effects. It makes a difference, also, to body cooling whether the wind movement is directed onto the back or onto the face—the latter having much more influence on body comfort or discomfort.

The reader should now come to appreciate the fact that relating building design to environment is definitely not a simple procedure. During the past decade a completely new science of building climatology has been formulated. Problems of domestic heating and cooling are analyzed, using the natural elements blended with artificial aids to achieve maximum economy and comfort. While designers and home builders continue their relentless defacement of the world landscape— from the Middletown, U.S.A., Tract Development to Housing for the Urban Bantu, South Africa—the research student can locate only a score of constructive, counteracting influences coming from a few experimental laboratories throughout the world. But from these few agencies we can surely hope to achieve design data for our adequately climatized, low-cost, owner-built dwelling:

· At the Hot Climate Physiological Research Unit, Oshodi, Nigeria, Dr. Ladell is conducting valuable research on shading effects.

· At the Graduate School of Architecture, Columbia University, a research group was organized in 1951 to study the influence of climate on the Macroform (general planning area) and the Microform (architectural details). To date they have made significant progress in the study of solar control and natural air conditioning.

· In Stockholm at the Swedish Institute of Technology, Gunnar Pleijel has published extremely interesting material on the use of the cold sink of the north sky as a cooling means. Cold spots in the north sky have been scientifically determined and their temperatures accurately measured. The reflection of the north sky against a wall has an effective sky temperature of 75°F, which is 45° lower than the average of about 120° for the south sky. The mirrorlike reflection of the cooler north sky explains why livestock will stand in the north-side shade of high-walled buildings in preference to proffered, conventional, over-head shades. Professor Pleijel's more recent work involves studies in natural lighting and window protection—protection, that is, from heat losses from inside the building and protection from unwanted heat gain from the outside.

· Architect Jacques Couelle, director of the Centre de Recherches des Structures Naturelles in Paris, has built a number of low-cost, naturally air-conditioned houses in Morocco. Ground-tempered air is channeled through an inside air space and released at the opposite end of the house.

· Dr. Ernst Schmidt, professor of thermodynamics at the University of Brunswick, has given considerable study to night-air cooling. His work has special value for use in desert locations where electric current is not available for refrigeration.

· At Forman Christian College, Lahore, Pakistan, Professor W. C. Thoburn built several experimental cottages which use a subterranean-tempered air conditioning. In one building, outside air is drawn into the windows of the cellar and then down an air-well to a 14-foot-deep underground tunnel which makes a rectangular circuit of 120 feet of running length. Air is pulled up through a central duct by means of a low-power fan and is distributed into each room above. This system of "lithosphere building" proved to be especially efficient for summer cooling, as the earth temperature at 15 feet below the ground tends to remain constant throughout the year ($76°F$ at Lahore).

· Wendell Thomas experimented with a simplified version of lithosphere building in two different houses in Celo Community, Celo, North Carolina. The basement, in one version, and the crawl space, in another version, provide ground-tempering contact as well as unique, air-circulating duct systems. Cold air from the exterior house walls circulates into the basement or into the crawl space through slots between walls and floor. The air is warmed by contact with the lower ground level (in the basement or in the crawl space) and then rises through a grille located in the center of the house to the actual fire chamber on the floor above. Both houses are sun-heated, and the house with the crawl space is, additionally, protected from heat and cold by a window sill height earth berm. In this house, without overnight artificial heat, the temperature seldom falls below $60°F$ on cold winter mornings.

· At El Rito, New Mexico, Peter van Dresser has spent much time and energy developing a low-cost, solar-heating installation. His own solar house engages a complete heat-collecting and heat-storing system, but in more recent work he is perfecting a partial sun-tempered system.

A. HEAVY CONSTRUCTION PROVIDES COOL CONDITIONS INDOORS DURING HOT DAYS — FOR DAYTIME LIVING.

B. TALL & NARROW WINDOWS — NORTH

C. ADJUSTABLE WINDOW SHADES ON OUTSIDE OF WINDOW. SHADE AND PROTECTION FROM DAY-TIME HEAT AND GLARE.

D. LIGHTWEIGHT CONSTRUCTION HEATS RAPIDLY AND HOT DURING DAY, BUT COOLS RAPIDLY AT NIGHT FOR SLEEP-ING AREA.

E. SHADE RIB ON WEST WALL PREVENT SUN ON NORTH WALL IN SUMMER EVENINGS.

F. LARGE OPENINGS ALLOW FREE AIR FLOW

G. WIDE OPENINGS ALLOW NO DEAD-AIR POCKETS.

H. HEAVY WEIGHT MATERIALS ON EAST AND WEST WALLS RESTRICT THE FLOW OF SOLAR HEAT DURING DAY.

I. LIGHTWEIGHT MATERIALS ON NORTH & SOUTH SHADED WALLS ALLOW RAPID COOLING AT NIGHT.

J. ELEVATED FLOOR PROTECTS AGAINST DAMP, PROVIDES BETTER AIR CIRCULATION.

K. LARGE GLASS AREAS ON SOUTH & EAST WALLS.

L. CENTRAL HEAT SOURCE FOR EVEN DISTRIBUTION.

M. WEST WALL PROTECTION FROM AFTERNOON SUN. HEAVY INSULATION NORTH WALL.

N. EVERGREEN SHRUBS OBSTRUCT WINTER WINDS. IN SUMMER BREEZES INCREASED BY CHANNEL-ED INTO OUTDOOR TERRACE.

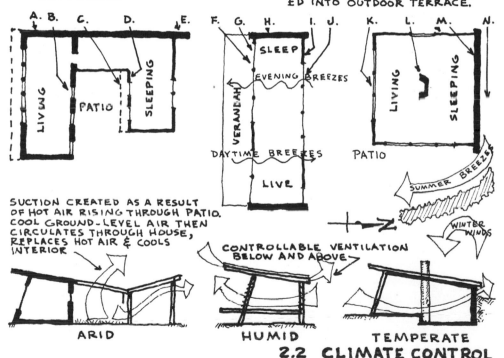

SUCTION CREATED AS A RESULT OF HOT AIR RISING THROUGH PATIO. COOL GROUND-LEVEL AIR THEN CIRCULATES THROUGH HOUSE, REPLACES HOT AIR & COOLS INTERIOR

ARID   HUMID   TEMPERATE

2.2 CLIMATE CONTROL

Throughout the humid, the arid, and the temperate regions of the world these (and many more) independent investigators are making their efforts known to all who will take the trouble to search them out. Though most of their research is still in its experimental stages enough can be learned by the individual home builder to be of great assistance to the owner-builder's planning a more economical and comfortable home.

In effect, this new science of building climatology is directed toward the control of climate. The term "climate control" is often seen in the literature on this subject. This control is managed in two ways: with artificial aids and/or through constructional means. Cooling by evaporation of water or by fans and warming by heaters and fireplaces are artificial aids, which are a part of the building. Items of basic equipment which do not call for consumption of fuel or power are considered "constructional." (Both methods of cooling and heating climate control will be discussed in the next two chapters.) From a practical, economic, or aesthetic point of view it makes much more sense to develop, where possible, constructional features for warming or cooling the owner-built home.

As a general summary of basic constructional considerations there is presented on p. 23 a model plan for each of the three climatic regions in the United States mainland.

3 VENTILATION

Enough is known today about natural ventilation for summer cooling to warrant the entire replacement of artificial air-conditioning devices. Stated in simplest terms (with reference to Figure 2.1), prevailing breezes enter small, louvered openings at the lower section of the south wall of a building, circulate across the "living zone," rise, and then exit through larger, higher openings in the opposite wall. What actually takes place will be described below. At this juncture, one only need know that the ventilation process is simple, dependable, and very important to the comfort and well-being of the house's occupants.

One might wonder to what extent an improperly designed window placement—or any other aspect of building design, for that matter—is responsible for social and domestic misery. Architect Frank Lloyd Wright once declared that he could design a house which would cause a couple to seek a divorce. Even such a house would, doubtless, be superior in design to the type of dwelling almost universally purchased or rented by average Americans, in which they unwittingly live "lives of quiet desperation."

Perhaps the greatest offense that house designers and builders commit against intelligent construction involves climate control. Most houses grossly violate the basic principles of natural summer cooling and sound winter heating. Ordinary rule-of-thumb builders are not the only violators of fundamental climate control principles. Architects of our largest and most modern hospitals, schools, and even skyscrapers fail to calculate such matter-of-course features as solar angles, ventilation effects, day lighting, and insulation requirements. The head research architect of Texas A. & M. University, Mr. William Caudill, an extremely capable designer, tells of boners he has made in his private building-design practice in the very area in which he specializes—namely, in the area of ventilation control.

A few years ago Caudill designed a magnificent school building in Texas. He took into account all the usual ventilation requirements for such a task; requirements for orienting the classrooms perpendicular to the prevailing breezes, allowing for openings to let air into and out of each room, for placing the rooms out of the range of wind-obstructing trees and buildings, and so forth. More than the usual amount of planning went into the provision to take advantage of the cooling effects of summer breezes. But, when the school building was finally occupied, teachers and students registered complaints about the excessively hot classrooms. Upon investigation, Caudill found that everything checked according to calculation. There was ample air flow through the classrooms. But, it was finally determined, the air was flowing along the ceilings instead of through the "living zone."

Obviously, the commonly used "architecturally projected" window used in this Texas school-building construction was ill-chosen for these quarters, inasmuch as fresh air was diverted to the ceiling. No other air diversion is possible with this type of window, although imaginative architects of another school building installed this same window upside down, achieving a successful downward flow of air into the living zone.

Basic to ventilation control is the principle that the direction and speed of air flow determine cooling effect. For air speeds of 200 feet per minute the cooling effect is equivalent to a lowering of dry bulb temperature by five degrees in still air.

Differences in pressure and in temperature are responsible for air movement through a building. When air enters a room at floor level and leaves through high windows it is the difference in air *temperature* which causes this movement. In many semitropical African homes cool night air is drawn through floor-level louvers extending the length of the house. Once inside, the air is warmed, rises and flows out through similar louvers near the ceiling. (When a ventilation system of this kind is employed, windows can be fixed and screens eliminated with the result that more light and a better view is realized.) This stack effect occurs in hot, dry climates where the temperature is appreciably higher inside the building at night than it is outside. In warm, humid climates where the temperature of the air inside the building is generally about the same as that outside, the stack effect is negligible.

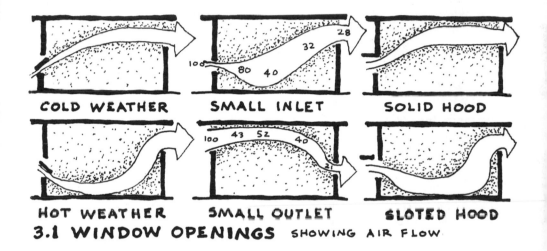

COLD WEATHER   SMALL INLET   SOLID HOOD

HOT WEATHER   SMALL OUTLET   SLOTED HOOD

3.1 WINDOW OPENINGS SHOWING AIR FLOW

Pressure differences result from the exterior collision of moving air against the walls of the building. When wind hits a building it piles up and then sweeps around the sides. A low-pressure area is, thereby, created on the side walls and on the leeward walls of the building, in contrast to the high-pressure area created on the windward wall. In terms of natural ventilation, it is the low-pressure area that is strategic. The low-pressure wall can be designed to suck air rapidly through the building. From a practical standpoint, this low-pressure suction wall should be designed with a much larger opening than that of the high-pressure windward wall. This basic principle of natural ventilation is quite contrary to common practice which, invariably, provides the maximum opening on the windward side of a building.

Wind tunnel tests at the University of Texas and at the South Africa Building Research Station have given us some very interesting ventilation facts. In one test, researchers were baffled when they realized that overhangs prevented natural air movement. But, when slots were provided in the overhang the desired ventilation was achieved. The slots, apparently, equalized the pressures from above and below.

Various window types influence air movement decisively, as will be illustrated in more detail in a following chapter on window design. Casement windows have the advantage that sashes can be swung clear of openings and can be adjusted to serve as wind scoops when breezes blow obliquely to the wall in which they are installed. Louvered windows are desirable for their provision of greater effective ventilation areas per unit of opening. Other types, such as double-hung and projected windows, partially obstruct the available opening.

Ideally, ventilation openings should be located in external walls, opposite each other. Air flow is greatest when the wind direction is within 30 degrees of normal to the opening. Beyond this angle, the flow decreases rapidly. Substantial screening reduces air movement, especially at low velocities. For example, 16-mesh mosquito gauze has been found to reduce air movement through an opening by about 60 percent when the wind speed is 1.5 miles per hour, but by only 30 percent when the wind velocity increases to 4 miles per hour. Exhaustive research at building research stations in Australia and in South Africa proved conclusively that the high-ceiling room, in itself, does not provide an increase in summer comfort. The traditional house in India has a ceiling of from 12 to 14 feet in height, but ventilation experts assure us that a 7 to 8 foot ceiling provides equal summer comfort, when other pertinent ventilation factors are observed.

In the eastern United States, summer relief depends primarily on air movement. Architectural schemes for ventilating interiors where no breeze can be counted on offer some of the most exciting prospects for low-cost summer cooling. A totally new concept of building design evolves from the application of basic aerodynamic phenomena. Ventilation by installation of windows becomes as obsolete as has the installation of shutters. Air can be brought into the building from the roof or from under the floor. Wind scoops can be designed to assist the flow of air from practically any angle.

In many tropical countries, native housing intuitively employs natural air ventilation. These buildings exemplify the "understanding"

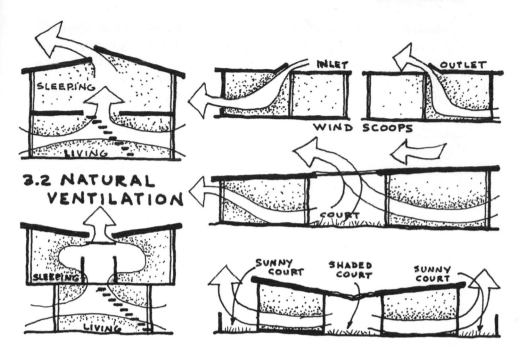

SLEEPING

LIVING

3.2 NATURAL VENTILATION

INLET

WIND SCOOPS

OUTLET

COURT

SLEEPING

LIVING

SUNNY COURT   SHADED COURT   SUNNY COURT

that upward convection of air, caused by differences in temperature, creates draft. A high-pressure area occurs at floor level and taps sources of cooler air, which flows in from outside through low vents. A sun-heated roof, on the other hand, creates a low-pressure area toward which air flows, rising naturally through centrally located clearstory openings, spontaneously drawing upward cool air from below. In order to function properly, all other openings besides the cool-air intake and the warm-air escape are closed during periods of calm.

The butterfly roof, which is the roof form best known for sending self-induced wind through a building, is a slightly more elaborate version of the natural ventilation intuitively applied by people of the tropics. If the lower-level of the unit is partly sunken and amply shaded by a cantilevered upper section, incoming air will be cooler by virtue of this increased shade and convection flow than will a unit without these features. With roof-ridge clearstory louvers open, a convection air current flows from the cool, shadowed, lower level through the house and out through the roof louvers.

The traditional patio court of arid regions can be adapted to supply substantial air movement through the house in periods of calm. As warm air rises through the roof opening, it creates a low-pressure suction. This hot air is replaced by ground-level air, which has been cooled by tree shade and lawn. If a treeless, sunny court is employed in conjunction with a shaded court an even greater natural air flow results. The shaded court forms a high-pressure zone, drawing air through the house from the opposite, low-pressure, sunny court. As this moving air passes through the house it ventilates inside rooms.

There is one area in every conventional house that should, but often doesn't, utilize natural air-flow ventilation—the attic. Perhaps one such house in a thousand has a truly sufficient development of attic ventilation devoted to summer cooling. Impartial investigations show that, if attic temperatures are to be sufficiently reduced by using gravity air-flow, the ventilation opening must be about six times that specified by current building-code requirements. Attic ventilation cannot be figured on the square-inch area of the opening in its relation to the attic's square-footage. Rather, it is the difference in the heights of the low inlet vents compared to the high outlet vents which makes for proper ventilation flow. In the placement of a majority of attic vents, this height difference is either negligibly provided for or, what is more usual, is not provided for at all.

Of the perhaps half-dozen methods for ventilating attics, the usual practice of installing wood louvers at gable ends is the poorest for its inadequate provision of effective ventilation. On the other hand,

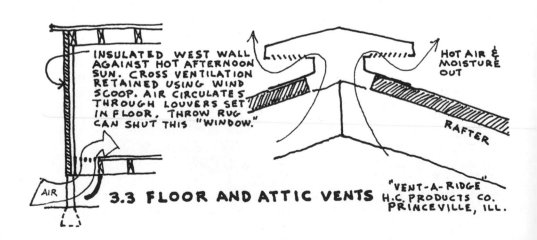

INSULATED WEST WALL AGAINST HOT AFTERNOON SUN. CROSS VENTILATION RETAINED USING WIND SCOOP. AIR CIRCULATES THROUGH LOUVERS SET IN FLOOR. THROW RUG CAN SHUT THIS "WINDOW."

HOT AIR & MOISTURE OUT

RAFTER

AIR

3.3 FLOOR AND ATTIC VENTS

"VENT-A-RIDGE" H.C. PRODUCTS CO. PRINCEVILLE, ILL.

the continuous ridge vent, with its provision for greater stack height, is the most efficient method for exhausting attic space, even though it utilizes an opening smaller than that in the louvered gable. Soffit and rake vents are becoming popular among progressive builders. Ventilation screens, located near the outer edge of the roof overhang, prevent wind-driven rain from penetrating to inside walls. High stack effects can also be accomplished by incorporating a flue-vent arrangement into the fireplace chimney. The mechanical, rotating type of roof vent seen atop industrial structures can be used to advantage for building aeration once its aesthetic limitation has been overcome.

Aesthetic limitation occurs in every phase of the building design and construction process. No major advances will be possible until such phenomena as gravity air flow, air-scoop ventilation, and warm-air convection become more a part of our working vocabulary, and until our aesthetic appreciation becomes more functional.

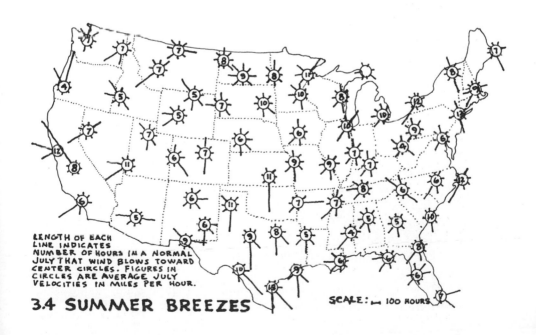

LENGTH OF EACH
LINE INDICATES
NUMBER OF HOURS IN A NORMAL
JULY THAT WIND BLOWS TOWARD
CENTER CIRCLES. FIGURES IN
CIRCLES ARE AVERAGE JULY
VELOCITIES IN MILES PER HOUR.

3.4 SUMMER BREEZES

SCALE: ⊢ 100 HOURS

4 SUMMER COOLING

The same good sense design, which is the feature of the naturally ventilated house illustrated in the previous chapter, can be incorporated into a naturally air-conditioned interior environment. Again, the starting point for this consideration begins where primitive, simplistic thinking left off centuries ago. We now know that systems for water cooling go all the way back to 2500 B.C., when the Pharoahs of Egypt employed slaves to fan air over large, porous earthen jars filled with water. Water, seeping through jar walls, resulted in the exposure of a large, wetted surface, which furnished evaporative cooling. Early American Indians of the Southwest achieved similar results with water-filled, porous earthen jars placed in open doorways. Water, seeping through the jars, quickly evaporated in the dry air, cooling both the water remaining in the jar and the surrounding air, which, subsequently, blew into the houses. The same system, called "olla," is today employed in Mexico, using porous earthen jars.

Pools and fountains cool effectively when used indoors. In Iran, for instance, it is common to find pools of running water inside a house, which is, itself, built semiunderground to escape solar penetration. Ventilating towers are provided overhead to catch wind from above

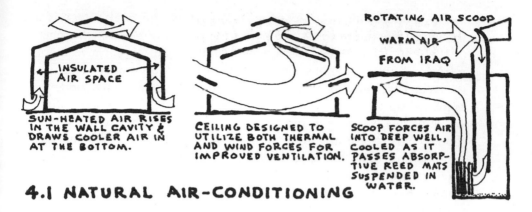

INSULATED
AIR SPACE

SUN-HEATED AIR RISES
IN THE WALL CAVITY &
DRAWS COOLER AIR IN
AT THE BOTTOM.

CEILING DESIGNED TO
UTILIZE BOTH THERMAL
AND WIND FORCES FOR
IMPROVED VENTILATION.

ROTATING AIR SCOOP

WARM AIR

FROM IRAQ

SCOOP FORCES AIR
INTO DEEP WELL,
COOLED AS IT
PASSES ABSORP-
TIVE REED MATS
SUSPENDED IN
WATER.

4.1 NATURAL AIR-CONDITIONING

the earth and to divert it across the water pool into the underground room. It must have been in Iran that architect Frank Lloyd Wright got his brilliant idea for cooling a house he designed in Mexico. During the summer months, the sunken fireplace hearth is filled with water. Down-draft air movement from the chimney circulates over this pool to cool the living room. This unusual fireplace has a summer-cooling as well as a winter-heating function.

A German patent is out for a system of storing cool air in cellars. The cooling capacity of the cellar is increased by filling it with crates of crushed rock, which possesses an enormous capacity for absorbing heat. Air, passing through the heat-absorbing rock, is cooled and fed by a powerful fan to the upstairs during the day, while the same fan at night brings in cool air after sunset.

There are important practical principles to be learned from indigenous and from primitive cultures. We find that the Arab's tent has slowly evolved throughout the centuries. As a functional shelter form, it reflects the hottest sun on earth. The Arab tent actually consists of two separate tents. The outer tent is white and acts as a heat-reflective layer. The lower, inner tent additionally protects tent occupants by providing a blanket of moving air between the two tent layers. As the most basic of all shelter forms, this tent system also illustrates the two, most basic principles of summer cooling: reflective insulation and ventilation. Without ventilation the effect of insulation can be completely nullified and proven inoperative. Insulation and ventilation systems must be designed with the respective cooling effects of each in mind. One system must supplement and reinforce the other.

The modern, air-conditioned counterpart of the Arab's tent is a thousand times more costly and performs at a fraction of the efficiency of the tent. It is, indeed, an air-conditioned nightmare. Since 1952, refrigerant air cooling has come into its own in this country. This was the year that mass produced home air-conditioning units suddenly appeared on the market. Twenty companies sold 250 million dollars worth of this equipment and were forced to turn down 100,000 customers. Now there are over one hundred companies producing air-conditioning units for an increase in output of 500 percent. Loan agency figures indicate that 60 percent of the new homes in this country have some form of central air conditioner included in the plans. We hear more and more about heat pumps, combination heat-and-air-conditioning units, dehumidifiers, package split-system units, and evaporative coolers.

Costs run high for all this air-conditioned comfort. In the 1950s, a survey conducted by the National Association of Home Builders determined that the cost of installing an average-sized air-conditioner in an average-sized house came to $1,308. 1952 operating expenses exceeded $70 per summer season. The heat pump, avowed king of year-round air conditioners, costs from $2,500 and up to install. Today, these figures are probably doubled.

There is no intention here to discuss the possible advantage of one type of air conditioner over another. In this writer's opinion, they are all too expensive to consumer and to ecology, alike, and too inefficient and unnecessary when elementary rules of good planning are respected. The design of a residential cooling system does not have to meet the same physiological requirements as does the design of a *heating* system. Heating systems are required to produce an indoor temperature of approximately 68°F, regardless of the outside temperature. Cooling systems, on the other hand, need only supply a drop of from 10 to 20 degrees below that of the outside temperature. Mechanical air conditioners producing greater cooling differentials than those normally occurring outside are a real injury to health. According to the American Public Health Association, "Heavy sweating in a hot, outside temperature leaves moisture in the clothing, which greatly increases the possibility of chilling when the body is, subsequently, exposed to lower, indoor temperature."

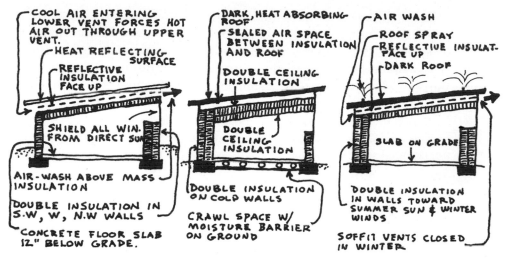

COOL AIR ENTERING
LOWER VENT FORCES HOT
AIR OUT THROUGH UPPER
VENT.
HEAT REFLECTING
SURFACE
REFLECTIVE
INSULATION
FACE UP
SHIELD ALL WIN.
FROM DIRECT SUN
AIR-WASH ABOVE MASS
INSULATION
DOUBLE INSULATION IN
S.W, W, N.W WALLS
CONCRETE FLOOR SLAB
12" BELOW GRADE.

DARK, HEAT ABSORBING
ROOF
SEALED AIR SPACE
BETWEEN INSULATION
AND ROOF
DOUBLE CEILING
INSULATION
DOUBLE
CEILING
INSULATION
DOUBLE INSULATION
ON COLD WALLS
CRAWL SPACE W/
MOISTURE BARRIER
ON GROUND

AIR WASH
ROOF SPRAY
REFLECTIVE INSULAT.
FACE UP
DARK ROOF
SLAB ON GRADE
DOUBLE INSULATION
IN WALLS TOWARD
SUMMER SUN & WINTER
WINDS
SOFFIT VENTS CLOSED
IN WINTER.

4.2 SUMMER HEAT—WINTER COLD—EXTREMES IN BOTH

The mechanistic approach to the satisfaction of heating and cooling requirements is based on the assumption of a steady flow of heated or of cooled air. Actually, however, heat flow varies with the time of day and with the seasons and is further influenced by site orientation, building structure, planting design, and the period of house occupancy. Heating and air-conditioning salesmen tend to oversimplify thermal requirements for homes, never mentioning the fact that heating and cooling needs are of a temporary and a local nature. Comparable with the use of electric fans and lights, artificial heating and cooling aids should be considered space equipment, that is, they should be employed only when and where they are needed. One of this country's foremost climatologists, Dr. Paul Siple, succinctly observed in a Building Research Advisory Board speech that, "although we have made constant improvements in buildings and have many mechanical developments to our credit, we must admit that some of these improvements, such as air conditioning, have really been developed in order to rectify errors or inadequacies in basic design." At the same conference, the feeling was expressed that our present technical attitude tends to produce buildings that are always "fighting their environment" instead of working with it. Buildings have become overengineered. Technical facilities are employed to counteract situations which common-sense attention to site, orientation, sun path, wind directions, and even to proper use of vegetation might well prevent.

Specialized, over-mechanization of modern houses has driven us into a vicious cycle of equipment acquisition and replacement, with no end in sight. For instance, about one-third of a ton of refrigerant is wasted to counteract the 3,400 Btu an hour heat alone generated by a TV set. This amounts to about $200 in first costs for an air conditioner. If one includes the heat generated by the refrigerator, water heater, and stove the first costs come to over $300—which is the money that is wasted before the rest of the room temperatures have been reduced by one-tenth of a degree.

There is only one alternative to this unending purchase of artificial aids to keep house interiors cool. A clear understanding of heat-flow principles, ventilation effects, and cooling requirements becomes paramount to this remedy. In some instances, a form of mechanical ventilation may be called for in a new home, but, for the most part, we can rely on natural air conditioning in conjunction with adequate and well-placed insulation.

It is a known fact that it costs three to five times as much to remove a Btu of heat from a house in the summer as it does to add one Btu of heat to the house in the winter. This statement can better be appreciated once it is realized that a black rooftop in any part of the United States on a clear summer day can be expected to reach 150°F. Roofs receive twice as much heat from the sun as do walls. Unventilated attics have been known to reach temperatures of 175°F. Our first and most obvious reaction to this fact can be summed up in one word: insulate. Insulation, yes, but which of the 150-odd brands of residential insulation should be applied? Should rigid insulating board, flexible blanket or batt, loose fill or reflective insulation be used? And—summer insulation is known to work in two different ways. It may keep much of the heat from entering the house through the roof, but the heat that does enter through the roof is trapped in the attic, remaining there and building up to reradiate onto sleeping occupants throughout the night. Bedroom insulation is unnecessary and restrictive of optimum summer sleeping comfort. A light, frame construction, which heats rapidly and is hot during the day, will also cool rapidly at night, offering better nighttime conditions than will a heavily insulated construction. In winter, the rapid response of lightweight construction to heating can be observed in the fact that, during sunny days, this type of construction is more readily warmed than is insulated construction. However, lightweight construction cannot retain heat well and is colder than insulated construction on winter mornings. In climates with hot summer conditions, daytime living areas, such as the living room, dining room, and kitchen, should be built of well-insulated construction to retard the rapid rise of temperatures indoors during the day. Evening living areas and bedrooms should be built of less-insulated construction, which permits more rapid cooling at night.

Compromise situations are inevitable when considering the various requirements for summer-daytime, summer-nighttime, and winter-daytime heating and cooling. We can, however, employ many natural air-conditioning factors, such as solar orientation, roof overhang, ventilation, selected window type, and shading devices. The National Association of Home Builders, at their air-conditioned village in Austin, Texas, showed that by shifting a house only 7° from a southerly to a westerly direction the benefit of a 36-inch roof overhang was practically nullified. In one case, the sun pouring through a large, unnecessarily exposed window boosted the heat load by 4,200 Btu an hour and increased air-conditioning operating costs by about 15 percent.

Roof pitch is another factor contributing greatly to the efficiency of insulation as well as to effective ventilation. Research experts assure us that a perfectly flat roof permits up to 50 percent more heat gain than a pitched roof on the same site. This illustrates the failure of flat-roof constrcution to get natural, hot air flow out from under the eaves.

As much as the exposed plank-and-beam, shed-type roof design is preferred by modern architects, it must be admitted that the gabled (attic) roof offers better resistance to heat build-up than does a flat-roof design. The plank-and-beam roof is difficult to insulate. Rigid insulation, applied directly under the roofing material on such a construction, restricts the passage of heat through the roof. Rooftop temperatures build up to the extreme and heat the tar of the built-up roofing to a flowing consistency, sliding the roofing right off the eaves. Ceiling insulation is, therefore, preferable to rooftop insulation. When it is realized that overhead insulation may save up to 90 percent of one's winter heat loss, it becomes obvious that the top of the living space is the main area to insulate for summer cooling as well.

Before the proper amount and the best type of insulation to install in one's house can be determined, one must have some idea of the regional requirements of his area for maximum winter heating and summer cooling. If one's house is to be built in an area of high summer heat-gain, one particular set of factors will influence insulation selection and installation. If one's house is to be built in a region that experiences extreme winter cooling, then another, entirely different set of conditions must be satisfied in order to maintain optimum comfort. In regions where extremes of heat and cold occur, a third set of factors may be operant.

Of the three ways in which heat flows through building materials— by convection, by conduction, and by radiation—it is radiation which contributes most significantly to summer heat build-up. On page three of the Bureau of Standards circular, *Thermal Insulation of Buildings*, the following passage appears:

> Although air is a very poor conductor of heat, the insulating value of an ordinary air space is rather small on account of the large transfer of heat by convection and radiation. Radiation is largely responsible for the ineffectiveness of air spaces bounded by ordinary building materials, such as are found in frame or other hollow walls. The low insulating value is often erroneously attributed to convection; but, as a matter of fact, from 50 to 80 percent of the heat transfer across air spaces of ordinary sizes takes place by radiation. If air spaces were bounded by bright metallic surfaces, the transfer of heat by radiation would be greatly diminished, since clean metallic surfaces are much poorer radiators than nonmetallic surfaces, such as brick, stone, glass, wood, paper, etc.

Reflective metal foil, either copper, aluminum, or steel, costs less than three cents a square foot and can be easily installed by the owner-builder. One should remember to place the shiny side out (or up) and to leave a $3/4$-inch, ventilated air space between the foil and the surface it faces.

Some home-building writers get overly excited about one particular type of material or about one particular system of construction which they recommend for universal usage, thereby performing a disservice to the owner-builder who cannot readily comprehend all of the principles governing its usage. Reference here is to the recommendation by Rex Roberts in *Your Engineered House* for the sole use of reflective foil for insulation. His multilayered insulation system is largely nullified by the fact that these layers are not separated by an essential, $\frac{3}{4}$-inch air space. Hard to improve upon are commercially produced, foil-faced fiberglass insulation batts, which ingeniously trap air in their generous matt of thin, glass fibers.

TABLE 4.1

ROOFING TEMPERATURES°

Color	Surface	Temp. °F	°F Cooling
Black (Asphaltic)	on wood	162	0
Black paint	on G. I.	161	1
Black plastic	film	157	5
Dark blue	roll roofing	154	8
Dark red	roll roofing	153	9
Alum.-asphalt paint	on wood	153	9
Alum. paint	on wood	150	12
Light blue paint	on wood	147	15
New galv. steel	(G. I.)	146	16
Alum. paint, 2 coats	on wood	143	19
Alum. spar enamel	on wood	142	20
White paint, dull	on wood	134	28
Alum. foil	on clear plastic	132	30
New alum. corrug.		131	31
White paint	on G. I.	125	37
White paint, shiny	on wood	117	45

° From bulletin issued by University of California, Davis

As much as 70 percent of the sun's heat rays can be reflected from one's house by the installation of a white or light-colored roof. A thin layer of quartz gravel or marble chip on a built-up tar-and-felt roof is by far the best type of surface for regions suffering a high incidence of summer heat. Where a water-cooled roof system is employed, a layer of porous, crushed-tile chat or water-absorbing, crushed coral is required. When water is used, the roof is cooled by evaporation as well as by reflection.

Experimentation with water-cooled roofs began in the late 1930s. At that time, experimental roofs were built perfectly flat and were flooded with about four inches of water. This system worked amazingly well during the day, reflecting about 80 percent of the sun's heat. But, during the night, the warmed water radiated heat down into the house. Mosquito breeding further disadvantaged the flooded-roof system. Moreover, the design had to include a roof structure strong enough to support the great weight of water.

More recent experiments with water-cooled roofs have conclusively demonstrated the effectiveness of water sprays. Ordinary, rotating lawn sprinklers operate quite satisfactorily when placed to give complete coverage of the roof area. Another method of roof cooling is the installation of a perforated pipe from which water trickles, cooling by evaporation and by reflection. Roof-spray tests were made at the Universities of Florida and Texas, where it was found that sprayed water reflects as much as 80 percent of the sun's heat, every pound of vaporized water removing 1,060 Btu of heat from the roof. The temperature of an experimental roof in the Texas tests was reduced 15°, from 132°F to 117°F, after spraying. Although it was found that the water thus consumed totaled 10 gallons a month for each square foot of roof, this amount was reduced by 50 percent when a thermostat, subsequently installed on the roof, was set to operate the system at 100°F.

Cooling systems of this kind require some mechanical means for moving large masses of air in at night and out during the day. The attic ventilating fan, or exhaust fan, is the least expensive, artificial cooling aid to install and to operate. A properly designed nighttime air-cooling system or daytime exhausting system, in combination with a correctly insulated and ventilated house shell, can replace the most expensive, mechanical air-conditioning unit. To make fan cooling comparable in results to air refrigeration, the fans must be correctly positioned in inlet and outlet openings. In an experiment conducted by Westinghouse Electric Corporation, a 16-inch portable fan, placed in a window, was used to suck air *into* a room. Air temperature was, however, only reduced 8°. The fan was then moved to a room location where it blew air *out* of the room. Temperatures fell 14°. The explanation for this is that when a fan exhausts air *toward* an outlet window it picks up additional air from the room. A larger amount of cool air thereby enters the room from an inlet window. For this reason a fan should be placed away from the window, at a distance

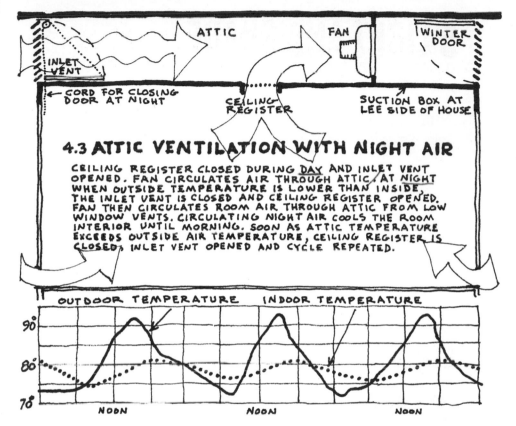

4.3 ATTIC VENTILATION WITH NIGHT AIR

CEILING REGISTER CLOSED DURING DAY AND INLET VENT
OPENED. FAN CIRCULATES AIR THROUGH ATTIC AT NIGHT
WHEN OUTSIDE TEMPERATURE IS LOWER THAN INSIDE.
THE INLET VENT IS CLOSED AND CEILING REGISTER OPENED.
FAN THEN CIRCULATES ROOM AIR THROUGH ATTIC FROM LOW
WINDOW VENTS. CIRCULATING NIGHT AIR COOLS THE ROOM
INTERIOR UNTIL MORNING. SOON AS ATTIC TEMPERATURE
EXCEEDS OUTSIDE AIR TEMPERATURE, CEILING REGISTER IS
CLOSED, INLET VENT OPENED AND CYCLE REPEATED.

that is twice the width of the window opening. The width of the
window or the opening should be twice the diameter of the fan blades.
A fan will eject 50 to 100 percent more air with such room placement than it would if placed directly inside the window frame itself.
Fans prove to be inefficient in regions where the relative humidity is
high, where there is little difference between daytime and nighttime temperature, and where night temperatures remain above 75°F.

The use of fans can solve the difficult problem of providing a high
U-value (or fast cooling) wall structure for summer and a low U-value
(or slow cooling) wall structure for winter. Cool night air can be
circulated when air is allowed to pass through vented wall studding
which is insulated with reflective foil of little mass. The air inlets are
arranged to be closed if so desired, making a dead-air space for insulation against winter cold.

5 LIGHT AND SHADE

Man's direct exposure to solar radiation is essential to his health. Sunlight deficiency, malillumination, is, like malnutrition, a very real malady, aggravating chronic illness, nervous disorders and the deficiency of Vitamin D and calcium. Some researchers, like John Ott (*Health and Light*, Devin-Adair, 1973), claim that altered growth occurs as a result of the blockage or the filtering out of ultraviolet light from our living environment. Mr. Ott says that the restriction of these beneficial solar wavelengths cause biochemical, hormonal deficiencies in animal cells, as well as in plant cells. He even suggests cancer-causing results in this neglect.

Early man's standard shelter was the cave. In this enduring and safe, though dark and damp place, the history of architecture begins and develops as man seeks to find new ways to admit more light into his dwelling. The history of building is in many ways the history of a struggle for light.

With the advent of the now-healthfully-suspect "miracle of glass" and with the creation of a skeletal structure that relieves the wall of its traditional load-bearing function, entire room areas can now be exposed to the outdoors. This attainment of interior light has been so popular the constructions of most contemporary builders have lost the quality of *darkness* which primitive man experienced in his cave. However, the skillful treatment of space, by architects such as Noguchi and Frank Lloyd Wright, preserves both the protected, cavelike, introverted space and the exposed, airy, extroverted space. At Wright's Taliesin West the cavelike living room contrasts with the adjacent airy music room under its light, canvas roof. A good designer will create both introverted and extroverted spatial experiences in his structures.

The usual punched-hole window openings fail to give us either an enclosed or an open feeling. The first American house built in wartime Java completely bewildered natives there. Instead of building walls of local bamboo, which is closely spaced to keep out rain while admitting light and air, the white man put up solid walls to keep out light and air, and then cut windows in the walls to admit light and air. Next, he put glass panes in the windows to admit the light but to keep out the air. Then, he covered the panes with blinds and curtains to keep out the light, too.

When planning the house for natural lighting, one should first distinguish between daylight and sunlight. Although both originate from the sun, sunlight comes to earth in a straight line while daylight is that light which is reflected or refracted from objects and substances that it strikes. Of the many factors known to influence sky brightness, some of these are the latitude and altitude of the locality, the time of year, the time of day, the amount of air pollution, and the relative humidity. Regional variations in natural lighting can be significant.

These facts are not mentioned in local building codes or in the guiding regulations of the F.H.A. or the Federal Public Housing Administration, which prescribe daylight illumination for a room merely on the basis of the ratio of glass area to floor area. One-tenth of the floor area is the generally approved standard for the required area of glass. Weather bureau maps, however, show that for the same latitude, the average daylight illumination is about 25 percent higher for the Plains states, the states lying between the Mississippi River and the Rocky Mountains, and is about 50 percent higher for the Plateau states, the states between the Rocky Mountains and the Sierra Nevada, than it is for the Eastern states. The natural illumination at noon in summertime Washington, D.C., is about twice that of this city in wintertime.

Rather than waste time complying with unintelligible building-code requirements, we should understand the nature of illumination with respect to the region of our locale and to the architectural effect that we seek to produce. Quality, as well as the quantity of light to be provided, should be our concern. Light entering a room, for instance, should be scattered by means of reflective surfaces. One should consider the degree of brightness, contrast and stimulation as it is to be experienced by the occupants of the room. By all means, conditions of glare should be avoided.

In our consideration of the need for glare control, we, rather startlingly, find that the customary practice of placing the kitchen or the bathroom window directly over the sink is ill-founded. This crucial window placement should more appropriately be located so that incoming light will arrive at the occupants' side or will illumine from above, not directly at them. Objects requiring our attention should be the brightest thing in view. When something else is brighter, such as the window in front of us, then an unconscious requirement is made of us to concentrate on that object which would command our attention. Our ability to effectively see the work at hand is thereby diminished, since our eyes will adjust to the brighter surface. Venetian blinds give excellent glare control. This type of window hanging reduces illumination from direct sunlight at the window location, and it increases illumination in the rear of the room by reflection from its slats.

Paradoxically, the larger the window installation, the less is the resulting glare. Contrast is eliminated between the window opening itself and the delineating wall around it. Shade trees planted at right angles to the windows and parallel to the chalkboard wall of a classroom will reduce interior glare. A continuation of the ceiling beyond the window in an external projection or an extension outward of the cross walls of the building, simulating vertical fins, will also help to reduce glare effects.

Light contrasts are especially decorative elements of design in the interior treatment of a house, but such contrasts should not be too forceful for the result will be tiring and sometimes bewildering. A gradual grading of color tone or brightness is desirable for effective, visual perception. Contrasting intensity between inside and outside light can be modified by curtains or drapes. A better solution to this problem of inside-outside light contrast would be the use of light-colored room decoration and the extension of the glass to the ends of the room to meet the internal return wall. This practice avoids a dark, shadowy frame around the window. Along the return wall, light dissipates evenly.

A score of variables contribute to our lack of information on window arrangement. Those variables include ventilation control, shading, light distribution, glare control, contrasting light intensities, and even the psychological effects of glassed-in areas. Definite psychological objections to large, glassed areas exist and should be recognized. Windows over four feet in width, those wider than the reach of the outstretched arms of a falling person, can induce fear for personal safety concomitant with the fear of the breaking and the shattering of glass. A horizontal structural member dividing the glass at a line somewhere between the height of the chair seat and the top of the chair back will afford one the assurance that his chair will not be accidentally punched through a floor-to-ceiling window.

Design for natural lighting should locate window headers as high as is practical. The intensity of daylight in a room on surfaces remote from the window increases as the height of the window header is increased. In fact, daylight penetration into a room is improved in greater degree by raising the header of the window in contrast to widening the window by the same amount of increase. A long, low window gives poor light penetration, while a very high window allows far deeper light penetration. This is partly due to the fact that the overhead sky is brighter than is the sky at the horizon. Without doubt, skylights provide the best possible source of natural lighting. Illumination from a corner window is high, but only near that window. Daylight penetration into a room from a corner window is actually quite poor, as is also the case with the light penetration from a bay window. Windows in more than one wall of a room will, of course, tend to admit and to distribute more daylight.

The traditional, three-fold function of the window, that of light, view, and ventilation, changes as modern technology develops newer, artificial lighting techniques and ventilating systems. Though modern structures allow great expanses of glassed (view) areas, there are better ways to provide lighting and ventilation than by using the opening window. For example, a fixed-glass window, accompanied by glass- or wood-louvered ventilation, has many advantages over the conventional window opening. An unobstructed view through horizontal frames thereby becomes possible. By keeping the glass area free of screening, vision is not impaired, and the glass does not get soiled, since screens are well-known to hold dirt for the first rain to spread on the pane.

Double glass panels restore some of the insulation lost when a window takes the place of a wall. Factory-made double glass panels, commercially called Twindow or Thermopane, are effective so long as the air-seal between the units is not broken. When these double panes are produced, only 15 percent of the air is left between the panes, and, if the glass or the seal cracks, the inside space will fog. These commercial units are unreasonably expensive.

Building-innovator Wendell Thomas of Celo Community, North Carolina, has devised a simple, inexpensive system for double- and triple-glazing window areas. The inside pane, preferably double-strength, is thoroughly sealed to the framing members with caulking compound and wood strips. In the fall of the year, the outside of this pane is cleaned, and a second pane is placed against it, separated by

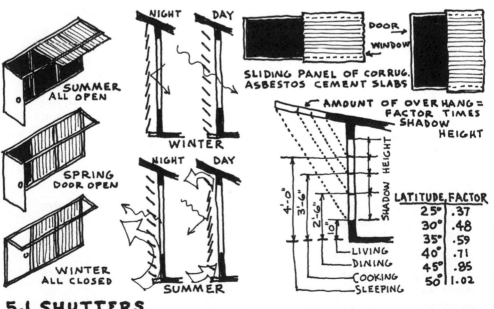

SUMMER ALL OPEN

SPRING DOOR OPEN

WINTER ALL CLOSED

NIGHT DAY

WINTER

NIGHT DAY

SUMMER

SLIDING PANEL OF CORRUG. ASBESTOS CEMENT SLABS

DOOR
WINDOW

AMOUNT OF OVERHANG = FACTOR TIMES SHADOW HEIGHT

SHADOW HEIGHT

LIVING
DINING
COOKING
SLEEPING

LATITUDE	FACTOR
25°	.37
30°	.48
35°	.59
40°	.71
45°	.85
50°	1.02

5.1 SHUTTERS

wood strips about $\frac{1}{2}$ inch wide. In very cold regions, a third sheet of glass may also be installed. It is important to keep the outer panes unsealed, for the air between the panes must be of the same low humidity as the outside air to prevent foging. Before the hot season starts, Thomas merely removes the few blocks around the outer glass panes and stores them until the following fall. Glass panels can also be designed to slide out of the way into adjoining walls. Inexpensive roller-guide hardware is available for this purpose. Shutters and sliding nonglass panels can further be used for window protection.

Full thermal insulation specifications for all-electric houses require a window sash "that provides no continuous metallic path from the inside of the structure to the outside air." Wood sash is, of course, a poor conductor of heat. It is, indeed, a shockingly 1,770-times-better insulator than is the currently popular metal sash. Wood, moreover, provides an effective barrier against condensation in its ability to maintain a stable temperature in contrast to fluctuations between inside and outside temperatures.

Tests for wood- and metal-sashed windows were recently conducted by the Forest Products Laboratory. Windows of the same size were exposed to the same temperatures. With an outdoor temperature of 20°, an exterior wood sash measured a warm 59° whereas metal sash measured a chilly 32°.

In the same tests about 20 percent more heat is lost through metal-frame windows. The wood-sashed windows remained free of condensation when outside temperatures dropped to 30° below zero, whereas condensation formed on metal-sashed windows in temperatures of 20° above zero.

Accordingly, the Forest Products Laboratory developed a low-cost, wood-frame window unit with permanently mounted glass and a transom ventilating unit, which is located above or below the stationary glass. This unit is designed to fit into 24-inch stud spacing, thereby eliminating regular window-framing headers, "cripples," and extra studs. Glass in this unit is placed near the outside wall to reduce weathering of the sash.

One possible improvement to be incorporated in this low-cost window design would be the substitution of translucent plastic (or fiberglass) for the regular window glass. Glass essentially blocks the entry of ultraviolet rays which are especially beneficial for human eyes.

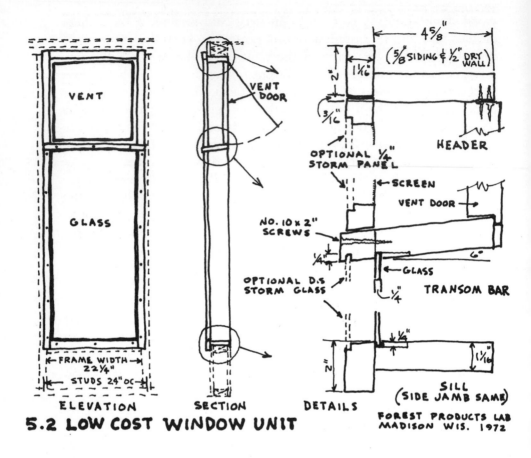

ELEVATION   SECTION   DETAILS

5.2 LOW COST WINDOW UNIT

FOREST PRODUCTS LAB
MADISON WIS. 1972

Plastic admits practically all of the sun's ultraviolet rays, while restricting the entry of infrared heat rays.

Windows and glass walls may be highly desirable from the standpoint of view, light, and ventilation, but they can create more problems than they solve if they are not intelligently planned. Glass, unfortunately, transmits 90 percent of the sun's heat and light. When sun heat enters a room through windows, the inside temperature is raised by 15 to 20 degrees. About one-fifth of all the heat that enters the average home comes through windows. It is a well-known but seldom-utilized fact that external shading is by far the most efficient method for controlling interior heat build-up. Solar radiation is blocked by this measure before it enters the house. An air wash

TABLE 5.1

RELATIVE EFFICIENCY OF DIFFERENT METHODS OF SHADING°

Method of shading	Description of Fenestration	Solar Heat Gain Expressed as Percentage of That Through Clear Glass

Special glasses and glazing materials	Corrugated polymethyl methacrylate sheets
	  a. Clear
	  b. Flat, white (translucent)
		100
		89

	Glass fiber sheets with plastic binder, translucent (corrugated)
	  a. Clear
	  b. Green
		83
		75

	Heat-absorbing glasses (tinted, laminated safety glasses)
	  a. Light tint
	  b. Dark tint
		80
		49

	Heat-reflecting glasses (thin, metallic films deposited between laminates) Medium tint	29
Double glazing	Ordinary clear glass (solar transmittance 0.86) both sides	86
	Heat-absorbing glass (solar transmittance 0.46) outside and ordinary glass (solar transmittance 0.80) inside	56
	Heat-reflecting glass (solar transmittance 0.11) outside and ordinary glass inside with air space ventilated to outside	17
	Ordinary glass both sides and white venetian blind in between	33

Internal shading	a. Net curtain with folds
	b. Heavy curtain with white lining with folds
	c. White roller blind
	d. Black roller blind
	e. Venetian blinds (closed)
	  White
	  Black
		75
		35
		36
		68
		
		43
		75

External shading	Awnings, louvers, etc., completely shading glass and allowing free air movement	20

° *by the National Research Institute, Pretoria, South Africa* between the shading screen and the window is also very effective. The screen should have a light-colored finish on the surface which is exposed to the sunlight. Table 5.1 illustrates the extensive range of heat-gain percentages which occur with various shading methods.

Inside shades are effective only to the extent that their reflectivity is adequate. The only real advantage of inside shading occurs in winter when heat retention is desirable or when the main function of the shading device is the control of day lighting and glare. Interior shades, such as roller shades, drapes, and insulation boards placed next to the window pane at night, will help to prevent heat losses through glazed openings.

Overhangs and horizontal sunshades above south-facing windows serve effectively as heat reducers. They also reduce glare. If maximum winter sun penetration is desirable while serving to shade the same area from summer heat, the extent of the overhang is a rather critical matter. Trellises for leafy vines, which will lose their foliage in the fall, should be set up over southern windows so that sunlight will penetrate into the house through the winter months.

An informative solar experiment in Pretoria, South Africa, conducted by the National Building Research Institute, showed that during the mid-afternoon on a summer's day the ratio of instantaneous heat gains through comparable glass and wall arrangements was about 7 to 1 on the north side, 3 to 1 on the east side, 11 to 1 on the south side and 40 to 1 on the sunlit west side. The best possible control for this situation is to use minimum areas of glazed openings, particularly on the north and west sides of a structure.

Another solar control device for east and west walls involves the use of sun shades. Quite a number of successful systems have been built, and, of course, there is the need for more research in this matter.

The reader should be reminded that four methods for excluding solar heat from buildings have been outlined and discussed: (1) *heat resistance*, using building materials which do not store heat from the sun, (2) *reflectivity*, maintaining light-colored surfaces on walls and roof, and using reflective insulation in ceilings, (3) *ventilation*, preventing reservoirs of hot air in wall cavities, and roof spaces, (4) *shading*, using overhanging eaves, sun-break devices, canopies, pergolas, and natural vegetation to shade hot walls and exposed window areas.

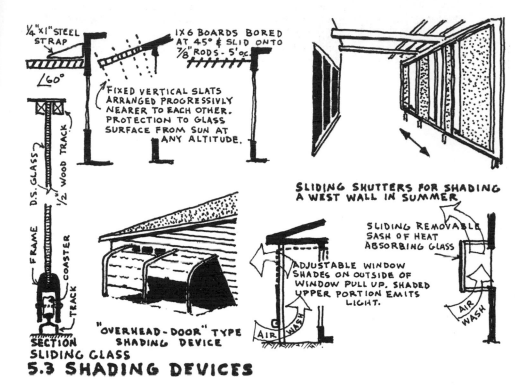

¼"×1" STEEL STRAP

∠60°

FRAME

D.S. GLASS

½" WOOD TRACK

TRACK COASTER

1×6 BOARDS BORED AT 45° & SLID ONTO ⅞" RODS - 5' o.c.

FIXED VERTICAL SLATS ARRANGED PROGRESSIVLY NEARER TO EACH OTHER. PROTECTION TO GLASS SURFACE FROM SUN AT ANY ALTITUDE.

SLIDING SHUTTERS FOR SHADING A WEST WALL IN SUMMER

SLIDING REMOVABLE SASH OF HEAT ABSORBING GLASS

ADJUSTABLE WINDOW SHADES ON OUTSIDE OF WINDOW PULL UP, SHADED UPPER PORTION EMITS LIGHT.

AIR WASH

"OVERHEAD - DOOR" TYPE SHADING DEVICE

AIR WASH

SECTION SLIDING GLASS

5.3 SHADING DEVICES

Designing for maximum summer heat control must employ all four of these methods, and each must be used so that the effect of one does not sacrifice the effect of another. For instance, a horizontal projection is installed to shade a window in summer. It is made of heat-resistant wooden louvers so that air-wash ventilation past the window is possible while, at the same time, improving air movement through the open window into the house. The louvers are painted white on both sides for greater reflectivity and for glare control.

We have, here in this example, a minor beginning in Integral Design—a "simultaneous approach" to the art of house planning. As will be elaborated upon in following sections of this book, the concept of Integral Design includes a consideration of structure and of the materials to be used and of planning for living functions and for the design of individual fixtures and appliances—as well as the inclusion of the exceedingly important climate and site-control features already discussed in this treatise.

6 SPACE HEAT

The National Bureau of Standards has defined a space heater as an "above-the-floor device for the direct heating of the space in which the device is located, without the use of external pipes or ducts as integral parts of such heating device." Of all the industrialized countries, space heating is more popular in England, where selective areas of the home are designed for a "background heat" of 50° from independent radiant heat sources. Our ideal in this country is, however, a heating system that provides uniformly comfortable 70° heat throughout the house, throughout the day. Essentially, this is the difference between space heating and central heating.

In rural areas of this country, there are about as many people who use space heaters as the sole source of heat as there are urban dwellers who use central heating installations. The low first-cost of oil-, wood-, coal-, or gas-burning space heaters accounts, in part, for their widespread use. This free-standing heater can be easily installed, and it is efficient as well as economical in its consumption of fuel. The provision for a zone of warmth by radiation, not possible with a central heating facility, is another important advantage of the space heater.

To appreciate the fundamentals of space heating, one must first arrive at an understanding of the Supplemental Concept of house warming. This concept involves the use of both multiduty appliances and of combination heating methods. In the former instance, it is possible to use a single appliance for space heating, water heating, and cooking. In England, this type of combination grate is becoming more and more popular. Such a space heater is designed with a side oven, or a top oven, and a hinged-closure hot plate for cooking. A boiler for domestic hot water is also integrated into the unit. Hot gases are directed around the oven or around the boiler by a single damper control. When the closure plate is up and the damper is closed, the effect is one of an open fireplace. A similar kind of British heating appliance is designed with the fire unit located in the living room and the cooking unit located in the kitchen, on the opposite side of a dividing wall. This type of back-to-back grate makes it possible for one appliance with one flue to serve two rooms. Since this type of appliance is not commercially available in North America, the owner-builder who wants one must build his own, using commercial iron grates and brick masonry.

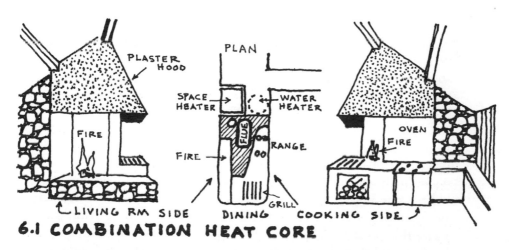

6.1 COMBINATION HEAT CORE

With the exception of the Franklin stove, the first improvement over the simple, wood-and-coal-burning iron stove occurred in 1836 with the invention of an air-tight radiant heater. It had a cylindrical fire box of rolled steel set between a base and a top of cast iron. Just twenty-five years ago, the next major improvement was made with the jacketed space heater, which had a sheet of metal enclosing an air space around the combustion chamber. Openings at the top and the bottom of this enclosed space, through which a current of air flowed upward, produced more converted heat than previous stoves, making it possible to heat rooms other than the one containing the heater. The most recent development in console, or jacketed, space heaters happens, also, to be the most significant. They are firebrick-lined, down-draft, automatically controlled, complete combustion heaters, such as the Riteway and the Ashley wood and coal burners. From a given amount of fuel these heaters are known to deliver fully double the amount of heat delivered by former styles of heaters. These highly efficient units will burn for the better part of a 24-hour period with one stoking. Combustion gases in the Riteway stove, distilled from the burning fuel in the heat chamber, flow down to the charcoal level and then pass upward through a special gas-combustion flue. There, preheated secondary air is added to insure complete combustion of all the gases. Nothing is wasted in this beautifully engineered system, since all that remains is fine ash and a flue relatively clean of wasted creosol tars.

Unlike the more sensible British combination-grate units, American space heaters have only the singular function of creating heat for room warmth. Furthermore, the only truly efficient heater manufactured in this country is exorbitantly priced and is, thus, beyond reach in cost to many low-income owner-builders. (Riteway Manufacturing Co., Box 6, Harrisonburg, Virginia 22801. No. 37 Heater, $288, F.O.B.)

Accordingly, the author has designed and built a wood-burning space heater from readily available, component parts. This unit follows principles of complete combustion and provides a cooking surface, water-heating coils, a heat exchanger, an oven, and a food dehydrator.

Free-standing, portable, console space heaters have been replaced in recent years by built-in varieties, such as circulating wall heaters, Panel-Ray radiant heaters, gas-steam radiators, electric-resistance heaters, and floor furnaces. Except for the fact that a floor furnace may cause excessive floor drafts, it is somewhat of an improvement over most space heaters. Cold air from the floor level, drawn into the furnace between the firebox and its jacket, moves upward to combine with the heat as it is emitted from the unit at only one point, through registers directly above the firebox. Combustion air drawn from the

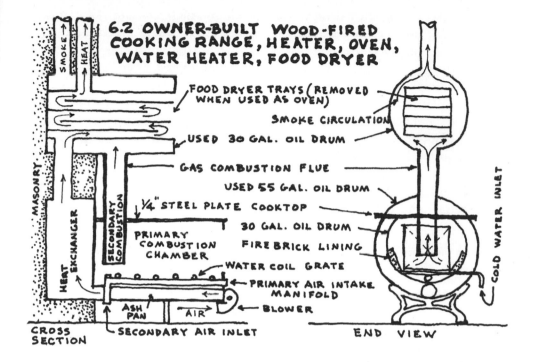

6.2 OWNER-BUILT WOOD-FIRED COOKING RANGE, HEATER, OVEN, WATER HEATER, FOOD DRYER

FOOD DRYER TRAYS (REMOVED WHEN USED AS OVEN)

SMOKE CIRCULATION

USED 30 GAL. OIL DRUM

GAS COMBUSTION FLUE

USED 55 GAL. OIL DRUM

¼" STEEL PLATE COOKTOP

30 GAL. OIL DRUM

FIRE BRICK LINING

WATER COIL GRATE

PRIMARY AIR INTAKE MANIFOLD

BLOWER

SECONDARY AIR INLET

SMOKE

HEAT

MASONRY

HEAT EXCHANGER

SECONDARY COMBUSTION

PRIMARY COMBUSTION CHAMBER

ASH PAN

AIR

COLD WATER INLET

CROSS SECTION

END VIEW

outside and the products of combustion are vented to the roof of the building. This furnace can supply heat to two rooms when placed directly under a stud wall with registers opening into both rooms.

In his own self-designed home in the mountains of western North Carolina, Wendell Thomas's heating system demonstrates one of the best examples of the Supplemental Concept of heating. The Thomases have not paid a cent for fuel. They burn only small quantities of brush in their Riteway heater. On the coldest winter mornings, without any overnight stove heat, the temperature inside this home seldom falls below 60°F. This high, minimum temperature depends upon four factors: (1) *good insulation,* including storm doors and windows and half-inch-thick insulation boards which are set inside the windows at sundown on cold evenings, (2) the chiefly *southern exposure* of the building, (3) the employment of the *cave principle* which banks earth up to window sill height on the south and the east and up to the roof on the solid-walled north and west, (4) the *"no-draft-floor"* invention which incorporates continuous, ventilating slots around the perimeter of the second story floor, where air, chilled by its passage over outside walls, sinks to the cellar or to a crawl space. There the air is heated by the earth which is up to 50° warm in the winter. It, then, rises through a central ventilator in the floor of the living space to be additionally heated as it passes around the heater stove.

Warmed air circulating through the ventilating slots between floors of the Thomas house not only tends to keep the living space warm, saving greatly on fuel, but it also tempers the air of both the upper and the lower stories of the house, keeping the cellar dry and the air of the living space normally humid. In the cool season, a constant, gentle, up-and-down air circulation is evidenced, even with all the doors and windows closed. Since air movement is vertical near the walls and near the heater and since it is only horizontal along the ceiling and over the cellar floor, there is no living space floor draft, no cold air sweeping from under the door to the heater, chilling the feet.

No-draft-floor principles are illustrated in Figure 2.1. Air-inlet slots are located around the perimeter of the house while outlets are centrally located at the heat source, a sunken-pit fireplace. Other design features, shown in this illustration, differ from those in the Thomas house. First, the sloping, south-facing glass wall allows a greater amount of winter sunlight to be transmitted to the inside. The amount of light transmitted through or reflected from the glass depends upon the angle that the light beam makes as it strikes the wall. In our hemisphere, the angle of incidence should not be less than 70°. Second, incoming solar-heat radiation is best collected and stored in a dark-colored masonry floor. Slate laid on a concrete slab is ideal. A continuous air space, provided below the cast concrete floor, will be discussed in a later chapter on floors.

Few builders are really aware of the potentialities for using earth-stored heat and cold to enhance residential comfort. By a process of combined conduction and absorption, heat is stored in the earth's crust every summer and is drained from it during the winter. At a depth of about 15 to 20 feet below the earth's surface, there is little variation from the annual mean. In India, investigations of this potential found that the amount of heat available from a 200-square-yard tunneled area was about four million Btu. An earth-tempered air stream was drawn from this passage 15 feet below grade through a system of masonry-lined ducts. Results from the experimental buildings designed to utilize some of this ground heat proved encouraging.

Heat engineers, in 1935, developed a mechanical device to withdraw and to capture the near-constant, inexhaustible heat from the earth. The system, called a heat pump, uses either an air or a water medium in its two-phase heating-and-cooling operation. After drawing heat-laden water from within the earth into a heat-exchanger refrigerant coil, the collected heat is concentrated by a compressor and sent to an air coil. Cool air from the house is blown through the air coil, where it picks up the heat and carries it through a central duct system. The heat pump in theory is a mechanical refrigerator in reverse. It extracts a quantity of heat from the ground or air, rejecting the remainder as unused energy. Thus, the heat pump has a low, overall efficiency, ranging from 8 to 30 percent. Until the time when heat-pump installation costs ($2,000 to $3,500) are reduced, they will not be competitive with existing central heating and cooling methods. Even though the electricity to operate a heat pump costs one-third as much as that of conventional electric heaters, the total operational cost of this device is still twice that which is paid for less expensive fuels.

The previously described Thomas house illustrates clearly the Supplemental Concept of house warming by combining several heating methods to achieve inside winter comfort. Of the three types of heating (wood burning, ground tempering, and solar rays) used in this home in the daytime when the sun shines, the south-oriented windows provide the Thomases with a major share of their winter heating requirements.

There are all degrees of solar heating, from simple, south-oriented, double-glazed windows to a complete collection-and-storage installation. Socrates was the first known writer to state the basic principle of solar heating, an important principle seldom employed in contemporary house-design practices. He wrote:

> In houses with a south aspect the sun's rays penetrate into the porticos in winter, but in summer the path of the sun is right over our heads and above the roofs, so that there is shade. If, then, this is the best arrangement, we should build the south side loftier, to get the winter sun, and the north side lower to keep out the cold winds.
>
> —Xenophon's *Memorabilia*

Socrates knew that south-facing walls in the northern hemisphere receive more radiation than other walls, but it has only been in recent years that we have known of the 4 to 5 times greater radiation received by the earth in its middle and higher latitudes from the low-angle winter sun. This fact, alone, makes winter solar heating a more viable consideration.

The provision for large, double-glazed, south-oriented windows with adequately designed overhang does not necessarily constitute a solar house. Even with a great deal of winter sun penetrating the house on clear or partly cloudy days, the heat loss during overcast conditions and at night may be greater than the gain during favorable periods. If no attempt is made to control excessive heat loss through glass, such a "solar" house may require as much as 20 percent more fuel than an orthodox house during December and January, according to *Heating and Ventilating Guide* figures.

Wendell Thomas recognized and solved this problem through the application of the Supplemental Concept of heating. He realized that, in the cold climate where he lives, a winter day is only eight hours long, one-third of 24 hours, and that the sun shines about every other day. So sunlight is available to him and his family only one-sixth of the time! Accordingly, he designed a dwelling only moderately solar and compensated for this by burying the house in the ground, except for the south and east window areas. Even here, the windows of this house do not fill the area, and, at night which is 16 hours long and on cold, dark days, the windows are covered from inside by insulation boards and by drapes, except for two large triple-paned windows, which can be left uncovered in the daytime to admit adequate light.

The Thomas solution is inexpensive and renounces the idea of the necessity for a predominantly solar house. Whether an inexpensive 50 to 100 percent solar house can be built is a question that research has not yet answered. There were less than thirty solar houses in the world in 1974. Every such solar house was either exceedingly expensive to build or required endless development hours from some well-heeled, professional gadgeteer. It is unlikely that solar-heating technology will ever trickle down to the low-cost, owner-builder ranks. If the fully solar-heated house ever becomes a viable alternative to the use of fossil fuels, it will very probably be a reality only for the upper-middle-class suburban dwellers.

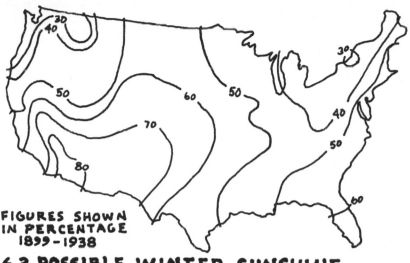

FIGURES SHOWN
IN PERCENTAGE
1899 - 1938

6.3 POSSIBLE WINTER SUNSHINE

The biggest obstacle to heating by the sun is the necessary capture of incoming solar rays and the storage of their energy. Obviously, there is need for some arrangement for storing this heat which is to be released at night or on dark days. Let's consider the general problems of building a house that is to be completely heated by solar energy.

There are good arguments for using income energy, abundant and inexhaustible sunlight, instead of using captial energy from fossil fuels and from uranium, which, once used, can never be replaced. But before solar energy can be utilized for house warming, it must first be collected and then stored. Both moving and fixed collectors have been used. Each type has construction and maintenance problems. The plane collector, which has advantages over the other types, consists of a black metal heat-absorbent plate, covered with insulating glass panes, which in turn enclose air spaces. Circulating air or water is used to carry heat from the black plate to a heat-storage bin.

Solar-heat storage is probably the major economic and operational liability of the solar-heated house. One must adopt a well-designed system of heat storage in order to maintain continuous service at standard temperatures during sunless hours. The specific-heat type of heat storage, using stones or hot water, was used in the first solar-heated houses built in 1939 at M.I.T. Later and more efficient storage installations of the heat-of-fusion type use material which melts and stores heat at moderate temperatures. The heat storage capacity of Glauber's Salt, one of the hydrates used for this purpose, is 6 or more times greater than the capacity of water and 13 or more times greater than the capacity of rocks.

In conjunction with the Supplemental Concept of house heating, however, the design for solar collection and storage takes on a different emphasis. When auxiliary heat is provided, the size of the collector plates, obviously, can be reduced, and the transport medium of water or of air can be integrated with the water-heating or the space-heating ducts. The economics of solar heating are improved when the system is designed as a "partial" rather than a "full" system. A house designed for between 50 to 75 percent solar heating seems to represent economically the optimum degree of solar heating possible for modest incomes. The term "sun-tempered" has been coined for this partial, solar heating.

One successful experiment with a sun-tempered heating installation at the University of Colorado cost under $500. A 12-by-20-foot heat trap, located on the roof of this installation, carried hot air, trapped at the eaves, through a double-glass panel, the bottom layer of which was painted black for heat absorption. From this collector, hot air was carried by ducts to the furnace system. Temperatures were varied by controlling the speed of the air moving through the layers of glass.

Peter van Dresser of El Rito, New Mexico, has worked for years to develop a simplified, less expensive, sun-tempered heat system. Below the masonry floor level of his house on an exposed slope, he has installed collector panels. Natural thermocirculation of hot air carries heat from the collector panels through ducts buried in stone under the floor slab. Correct tilt of the glass collector, which is latitude plus 15°, insures direct incidence of the winter's low-angle solar rays while high-angle summer rays are reflected away.

The William Johnston residence in Stratford, Ontario, Canada, employs a simplified sun-tempered heating system. For a slight additional expense in window and floor materials, the Johnstons' winter heating bill came to $97 in an area where similar but nonsolar-heated houses use up to $340 in fuel each season. Johnstons used the concrete slab which forms their floor to store heat coming in from windows.

The slab was poured over 6 inches of washed gravel and over 1 inch of insulating board, insulating the slab both from the outside walls as well as from the ground. The finished, concrete floor slab radiates heat directly to persons and to objects in the room. Its capacity to store heat is remarkable. It does not seem warm to the touch, yet it absorbs enough heat during the day to keep the house temperately warm throughout the night in all but the coldest weeks of winter. On one winter day when the ouside temperature was 15° below zero, it has been reported that the slab trapped and radiated enough heat to keep the house temperature at 76°F.

Successful sun-tempered houses in the Southwest have been built using sun-facing, black-surfaced masonry interior walls to store daytime solar energy. Surfaces inside such a house should be low absorbers and high emitters, such as finished plaster which absorbs 35 percent solar radiation and emits 93 percent radiation in long waves. Short solar waves, accompanied by long waves, penetrate glass readily and warm objects within a room. Long waves, however, cannot pass back through the glass. The resulting heat build-up is known as the Greenhouse Effect, and objects in the room, supplementarily warmed by the heat of captured long waves, radiate this heat to the building.

Of the more recent solar-heating developments around the world, perhaps none is more significant than the solar wall of a building erected by Trombe and Michel, in the French Pyrenees. Vertical, south-facing walls, having a high thermal storage capacity, are used

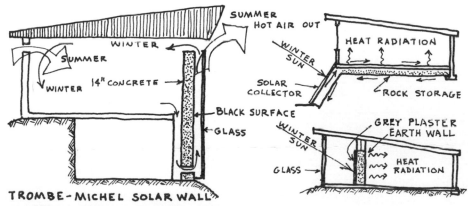

TROMBE-MICHEL SOLAR WALL

6.4 SUN-TEMPERED HEATING

as collectors of solar energy. A natural thermocirculation of air warms and, in the summer, cools the interior. Ducts allow cool air to flow in at the bottom of the wall and heated air to flow out of the top of the wall. In the summer, the collector acts as a thermal chimney. Hot air rises through it, pulling cool air from the opposite north wall through interior rooms.

Glass, however, is not required for the collection of solar heat. The National Physical Laboratory of Israel, Box 5192, Jerusalem, has perfected a highly polished metal surface coated with a molecular-thin black layer of special paint, which absorbs more than 90 percent sunlight. The polished metal radiates very little of the heat it receives through the black surface coating. These selective black-surfaced collector plates are widely used in Israel. Dr. Tabor, the laboratory director, has written, "We are at present [1960] negotiating with a large American company to license them to use our methods, but I cannot yet say what their policy will be towards individual builders and architects."

A major advance in sun-tempered space heating will probably take place when owner-builders can purchase selective black collector plates at a reasonable price. A south-facing wall fitted out with these plates would really "drink in" solar heat, windows or no windows.

7 CENTRAL HEAT

A pertinent first question often asked of a prospective home builder is, "How are you going to heat it?" The choice of the type of heating system for one's home has a major effect on room location and window placement, as well as on the general design of the house and its site-climate orientation. In answer to this question, most people have little more information than that which is inadequately offered by heating-appliance salesmen, heating contractors, and fuel distributors.

Heating concerns are complex. For instance, consider the operation of just one type of fuel-consuming appliance, the oil burner. When a high-pressure oil-burning unit, such as that produced by the Carlin Company, is fired for house heating, about one gallon of oil per hour is burned. A low-pressure burner, the Oil-O-Matic made by Williams, requires just half this amount of fuel per hour. And, the Iron Fireman Company has introduced a Vertical Rotary oil burner which uses about one-third gallon per hour less fuel than that model just previously mentioned.

Electric power companies widely advertise the advantages of the all-electric house, the freedom from handling fuel and its residues and the simplicity and flexibility of electricity usage. But, with electric rates at 3¢ per kilowatt-hour, heating costs are about 6 times as much as those using fuel oil at 16¢ a gallon.° Where natural gas is available, the cost differential is even greater. One should not, however, rule out the use of electricity for domestic heating. In regions where electrical rates are low or where mild winters prevail or in cases where an intermittent, quick-response type of heater is preferable, electricity may offer performance inducements regardless of cost.

The climate factors of air temperature, relative humidity, solar radiation, and air movement are, of course, of importance to heating engineering. It has been found that a wind of only 15 mph may increase the heat loss from a window surface by 47 percent or from a concrete wall by 34 percent. Therefore, heating plans have a critical relationship with windbreaks and with wind baffles.

Experiments at the Lake States Forest at Holdrege, NE, affirm the value of windbreaks. In these experiments, exact fuel requirements were recorded in two identical test houses. One house was exposed to the winds and one was protected by a nominal windbreak. With both houses maintained at a constant, 70° inside temperature, the house having the windbreak protection required 30 percent less fuel.

Heat sources are of three kinds: conductors (for example, warm floors), warm air, and radiant panels. Rather than attempt to provide warmth with one type of heater, the home builder might better be encouraged to combine the best features of several heating methods. It is certainly commendable to include in one's heating arrangements the radiative and conductive effects of both solar heat and the heat-circulating fireplace. Hot-air convectional heating, with its quick response feature, can, in another circumstance, compensate for the time lag typical of hot-water panel heating.

Ancient Romans and, much later, Count Rumford and Ben Franklin and, recently, heating engineers and physiologists have speculated about the effects of the heating process in relation to human health and to the economy of fuel. From all available current evidence, one is reasonably directed to use a radiative means of heating rather than a convected, warm-air type of heating. It is important to realize that the purpose of heating a building is not to put heat into the occupant

° Economy fluctuations necessitate substitution of recent, local prices for reliable comparison.

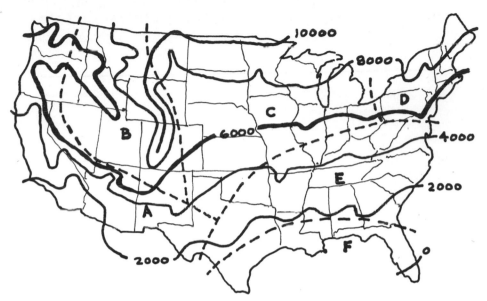

7.1 CLIMATE AND HOUSE HEATERS

A—mild winters
B—very cold winters
C—cold winters with wide
    temperature range

D—long cold winters
E—cool winters with cold
    snaps
F—short mild winters

CLIMATE AND HOUSE HEATERS

2000 degree-days*—intermittent types of heat: stoves, portable heaters using gas or electricity; central heat likely to be troublesome.

4000 degree-days—space heaters popular; baseboard radiation very satisfactory; electricity and bottled gas often used; heating limited to living room.

6000 degree-days—central heating desirable, though often replaced by space stoves; hot-water heating systems popular; electric heat impractical.

8000 degree-days—heat required in every room, preferably central system; periphery usually sufficient to permit use of

---

*The heating engineers' "degree-day" is based on difference between outside temperature and 65° F., counting by hours.

TABLE 7.1

COMPARATIVE COST IN DOLLARS OF DOMESTIC FUELS
(From *Consumers Bulletin*)

Fuel Type	Average Price°	Average Cost per million BTU
Anthracite Coal	$28.50 per ton	$1.40
Bituminous Coal	16.25 per ton	.80
No. 2 Fuel Oil	.16 per gal.	1.40
Manufactured Gas	1.40	3.35
Natural Gas	.75	.90
Bottled Gas (Propane)	.14	1.90
Electricity	.03 per KWH	8.80

° As there is a wide price variation from city to city, actual local prices should be substituted for reliable comparison.

but to keep him from losing heat. We are comfortable when we effortlessly give off heat, at the same time that we produce it. The only purpose of a heat system is to aid the body's mechanism, which maintains the balance between the rate of heat loss and the rate of heat generation.

With convected heat, the air temperature in a room must reach 68° to 70°F for basic comfort, yet this temperature, unfortunately, is too high for us to discharge heat rapidly. Our pores tend to open as a consequence of certain nervous and endocrine reactions, and blood flows to the surface of the body, where it radiates heat to the outside atmosphere. The result of high room-air temperature is akin to the feeling of exhaustion one experiences on a hot, humid day.

Comfort at lower air temperatures can be achieved by using radiative heating methods to promote the generation of heat within the body and to prompt the exhilaration that goes with brisk activity. In a conventionally heated room, hot air rises from the convector, usually located beneath a window, to sweep across the ceiling where it flows down the opposite wall. Temperatures are highest at the ceiling level, where they actually do the least good comfortwise. A temperature of 100°F at the ceiling may result in a mere 70° temperature at the lower, living zone. A smaller range of air temperatures between floor and ceiling is possible when using radiant-panel heating methods. Convected-air heating requires 70° air temperatures, whereas 65° are required using a radiative means of heating. The result is a 30 percent saving in fuel consumption.

Both the sun and an open fire broadcast radiant heat rays. The Romans, while in England 2,000 years ago, circulated hot gases from charcoal fires through ducts to warm walls and floors, actually discovering in the process what we now call "radiant heat." The traditional Korean ondol heating system adapts the radiant principle. Combustion gases from a kitchen stove flow through a labyrinth of ducts under a masonry floor slab to a chimney at the far end of the heated room. Radiant heat rises from the floor. Frank Lloyd Wright, in 1937, revived radiant heating in the Western world when he developed the gravity heat employed in his Johnson's Wax Building.

About 90 percent of the currently installed radiant panels use hot water as a circulating medium, but a radiant hot-*air* heating system is considerably less expensive to install and to operate. Most water systems use steel or copper pipe buried an inch or two beneath the top surface of a concrete floor slab. In order to achieve maximum efficiency, the water temperature in a radiant floor slab must be maintained in a range from 80° to 90°. Yet, a floor temperature of over 70° will cause a rise in foot temperature and create a disturbance in the normal heat emission from various areas of the body, since the temperature of the lower extremities is normally several degrees lower than the trunk and the upper reaches of the body.

The hot-water, radiant floor-heat installation is, no doubt, less costly than ceiling or wall installations, but the hot-water, heat-radiant ceiling has many factors in its favor. The fact that, in a ceiling-heated room, the floor will be warm may seem contradictory, but, if the entire surface of the ceiling is heated, there will be no convected air circulation, making even the floor appear to be warmed. This heat arrangement furnishes a uniform room temperature, with radiant energy being transmitted downward for interception by the surface of the floor. Water temperature in ceiling-heated surfaces must be kept from 125° to 135° degrees warm.

There is a lengthy time lag in radiant floor heating, involving slow morning heating and slow evening cooling, and a sudden change in weather cannot be compensated for quickly. This objection to a hot-water, radiant-heat system can be overcome by the installation of a thinner floor slab, which will hasten the response of the system to temperature changes. Dividing water circuits into separate sections—one circuit for the living area and one circuit for the sleeping area, etc.—will help to cut down on this heating lag. Also, a grid system of pipe layout is more efficient than a sinuous pattern.

The hot-water radiant ceiling will, of course, cut down the heat lag considerably. Proven to be far more efficient than the conventional plaster-lath ceiling surface is the latest development in ceiling heating—that of attaching the heat coils to the upper side of perforated, snap-on, metal panels. (Acoustical, thermal blanket insulation is part of the installation.) Metal is an excellent conductor of heat and will be heated to almost the same temperature as the water in the pipes. The exposed metal should be surfaced with a matt or flat finish, brushed or burnished aluminum, perhaps, being the best heat-conducting surface finish for the panels. If this metal panel is polished it will have no radiating quality.

Another recent development in hydronics (water heating) has resulted from experimental work at the University of Illinois. Considerable time and installation expense can be saved, it was found, by using $\frac{1}{4}$-inch to $\frac{3}{8}$-inch flexible copper tubing in place of the usual $\frac{3}{4}$-inch steel or copper pipe, since the number of required fittings can be reduced by one-half. A small, appliance-sized, automatically fed boiler has recently appeared on the market. High-temperature water heating for the floor combined with water heating for domestic use can be supplied in this manner at relatively low cost. Levitt, the mass tract builder, has used both $\frac{3}{8}$-inch copper tubing and this combination water-and-heating appliance (York-Shipley, 94,000 Btu an hour), in his house-heating system.

An hydronic system, known as radiant baseboard heat, is another current development. Heat issues from units located on cold walls. Some convection effects (drafts) occur, but these units are largely radiant in their effect. Forced, hot-water baseboard radiation is low in first cost and simple to install.

Fifty years ago, the cast-iron stove was moved into the basement of the house to become a furnace. As a gravity-supplied convection heater, it sent hot air, along with other gases (!), up through a grille in the floor to interior rooms. Basically, this warm-air gravity-heating system has not been improved. It is still the cheapest heating system for the small, compact home, and is, perhaps, found more often than

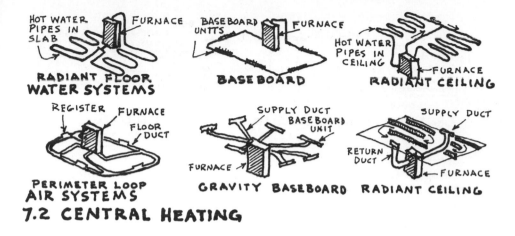

RADIANT FLOOR WATER SYSTEMS — HOT WATER PIPES IN SLAB · FURNACE

BASEBOARD — BASEBOARD UNITS · FURNACE

RADIANT CEILING — HOT WATER PIPES IN CEILING · FURNACE

PERIMETER LOOP AIR SYSTEMS — REGISTER · FURNACE · FLOOR DUCT

GRAVITY BASEBOARD — SUPPLY DUCT · BASEBOARD UNIT · FURNACE

RADIANT CEILING — SUPPLY DUCT · RETURN DUCT · FURNACE

7.2 CENTRAL HEATING

any other in homes across the country. Air enters the system through one or more cold-air or return-air registers and is heated as it passes through the large return duct in the furnace.

About twenty-five years ago, someone had the idea to install a fan in the bottom of this heater. The resulting forced-air system permitted smaller ducts and more freedom in design. Moreover, it was possible to arrange the house and the furnace on the same level. The majority of new homes are equipped with forced-air perimeter-ducted heaters or forced-air baseboard heaters, in spite of the fact that extensive research has proved that convected-air heating is generally unhealthful. High temperatures in this system burn or scorch the air, deoxygenating it.

The best method for domestic heating may be a wise combination of the radiative and the convective systems. Mention may be made here of two more promising combination systems of house heaters. From a centrally located furnace hot air is blown down into radiating feeder ducts imbedded in a concrete slab. Hot air, circulating through these feeder ducts, heats the floor to a temperature of about 70° as it passes to a larger perimeter duct and on into the room. Consequently, a blanket of warm air passes up the exterior wall where it is most needed. Since the floor surface is heated, no cold-air-return floor register is necessary. The absence of cold air at the floor level contributes to living-zone comfort.

Crawl-space perimeter heating is another recently developed com-

7.3 WARM AIR RADIANT-CONVECTOR

bination system. It is claimed that this system provides uniform temperatures with quick response at low cost. In this method, the total crawl space serves as the plenum. A central down-flow furnace supplies warm air to a short, stubbed-out duct system, which is aimed at the far corners of the house. Registers are located around outside wall perimeters. Return air is collected at an interior wall and returned to the furnace through a short duct. When a layer of heated air exists below the floor joists, not only is the temperature of the floor surface increased but temperature in the living zone is made more uniform from foot to breath level.

The National Fire Code Organization reacts against any heating system that permits moving, warm-air heat to be stored in wood flooring. This is one more instance where unrealistic code enforcement hampers improved, low-cost heating installation. In this system cold floors are eliminated as the floor, itself, becomes a heat duct. Warm air from an under-the-floor duct, a plenum, flows evenly around the cold perimeter of the house. Heat does not arrive in the building interior as a blast of hot air from a few registers, placed under windows. In many respects, hot-air registers are bad news. Such registers concentrate the heat flow into a room, creating hot, undesirable air currents and failing to provide anything like the even atmosphere of warmth achieved with radiant floor heating.

There is one ingenious method for overcoming the objections that the National Fire Code imposes on the hot-air plenum heating method. Wendell Thomas was on to a great idea when he found that, by providing a continuous, ventilating slot between the floors of a building and its outside walls, as described in chapter six, the living space could be kept adequately warm. It has also been found to be additionally desirable to place the heat source, such as a simple wood stove, away from the air duct, which leads into the plenum. The plenum will, then, circulate room-temperature air, not furnace-heated air. Air drawn from all parts of the house is mixed in the central duct and is then fan-driven into the plenum for redistribution. Warm areas are cooled and cool areas are warmed. Tests by the Department of Agriculture. Production Research Depot, No. 99, made on both the circulation plenum and the hot-air plenum, found both systems equally effective. The circulation plenum, however, has the additional advantage of offering a hot spot for rapid, body warm up by providing personal access to the space heater itself. This accommodation may seem exaggerated, but, when one comes from the cold into a modern, evenly heated house, the warming experience is neither immediately physically beneficial nor psychologically edifying. Every house needs a warming spot where persons coming in from the outside chill can, if for no other reason, warm hands and hearts.

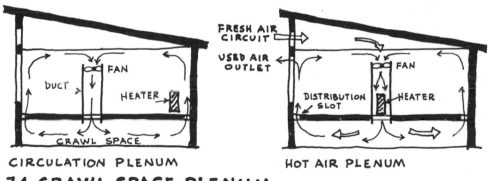

CIRCULATION PLENUM   HOT AIR PLENUM

7.4 CRAWL SPACE PLENUM

No matter what type of heating system one chooses, if the house is not adequately insulated and weather-stripped, heating costs will be major. In cold climates it will cost only half as much to heat a well insulated building as it will cost to heat a poorly insulated one. The Housing and Home Finance Agency reports that by moderately insulating a typical dwelling in Washington, D.C., the annual fuel saving will amortize in two years the additional cost expense of the insulation! In one carefully planned experiment, it was shown that, by coating the walls and ceiling of a construction with aluminum paper, the heating load was reduced 21 percent. It has also been found that, in a house with a concrete slab or with a crawl space, 80 percent of the hourly heat loss occurs through perimeter walls and through the floor. In two houses in Champaign, Illinois, one house, not weather proofed, required 3,000 gallons of heating oil in one year. The same sized house, with storm sash and storm doors, weather-stripped doors and windows, and insulated ceilings and sidewalls, consumed only 800 gallons of fuel. New types of weather-stripping, vapor barriers, storm sashes, and sheathing materials supply air-tight construction, requiring less heating fuel. As a result, today's modern, weather-resistant residences contain less moisture-holding atmosphere to plague home dwellers with the problems of condensation.

Inside buildings cold weather condensation results from humidity build-up. A family of four daily converts into water vapor an average of 3 gallons of water. This may be merely the result of ordinary household functions and normal human respiration. But, in addition to this expected humidity, many heating units have built-in humidifiers which further increase relative humidity and consequent condensation inside the house. A certain amount of low-moisture, outside air should, therefore, be brought into the structure through a cold-air return duct. This requirement becomes especially important to the effective functioning of a fireplace, when it is included as a part of the heating system, as will be seen in the following chapter.

8 FIREPLACE HEAT

Even at the risk of being labeled by some a male chauvinist, the author intuitively respects and personally prefers the Chinese representation of the house as feminine (Yin) and the house site as masculine (Yang). As a container, the house is hollow, womblike, commodious, and warm, and it is organized, managed, and cared-for by women. It is quiet, passive, and even submissive and contrasts markedly with the bright, forceful (male, if you will) elements of the open landscape about it. The fireplace is the one contrasting, male-dominant element within the house, and the most obviously phallic symbol is, of course, the fireplace chimney. The design of a house around its massive, central fireplace has, somehow, always felt right to this writer-builder. One may conjecture that all this discussion is merely an overindulgent, specious emphasis on the male concept, so my wife is moved to interject some words of caution. "The Chinese," she remarks, "speak of duality in all things. Just remember how forlorn is the chimney, standing alone after a house fire, with its support, the house about it, gone. And, after all, what is a home without a welcoming hearth?"

The fireplace designed and built by the author and illustrated on p. 74 probably represents considerable masculine symbolism. It is un-

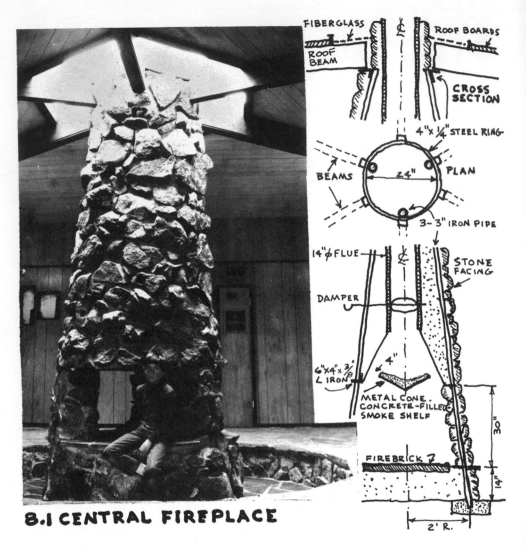

8.1 CENTRAL FIREPLACE

obstructed, free-standing, and insistent, at the same time that it is
supportive of the roof. It is central to the room in which it is placed
as it stands, sunken, in its fire pit. It is recessed in the dimmest area
of the enclosure, but, where the massive stone chimney rises to pierce
the roof, a translucent band cascades filtered, mellow illumination
down the stone face.

Man is the instrument of his symbol in his insistence on tending the fires—even if he does none of the other work in the house. Any observant architect will be a witness to the fact that it is, more often, the man of the house who is concerned with the design of the fireplace. An observant architect will also find that the woman client allows him more freedom with the landscape plan and the fireplace design than with any other component part of the house design. Clients approve contemporary line and form in their garden plan and with their fireplace form but appear to reject the same principles of good design when applied to the organization of the house proper. This fact gives a designer cause to expect that, functionally and aesthetically, he will realize better results with the fireplace form than with the plan of the room in which it is placed.

Unfortunately, most of the technical improvements in fireplace design have not yet emerged into common usage. This is so in spite of the fact that improvements occurred as far back as 1624, when Louis Savot invented the first heat-circulating fireplace. His unit was installed in the Louvre, Paris, and became the prototype for Ben Franklin's 1742 Pennsylvanian stove. The 1624 French fireplace achieved 30 to 45 percent more efficiency than do most American tract-home fireplaces of today! Savot surrounded the grate of his creation with a metal air chamber, which had warm air outlets above the fire opening. He also supplied the fire with air from under the floor. Thus, room drafts were reduced and combustion efficiency was further improved.

Few people are aware that practically all of the technical features of Franklin's Pennsylvanian stove were copied from earlier inventors. Savot's concept of a preheated draft was employed by Franklin with little change in design. Prince Rupert's descending flue, invented in 1687, was also applied on the early Franklin stoves. The smoke rose in front of a hollow metal back, passing over the top and down the opposite side. At the same level as the hearth, the smoke ascended the flue. Ducts, similar in design to those invented by Nicholas Gauger, in 1716, were also incorporated in the Franklin stove.

The most noteworthy development of the open fireplace took place in 1796, when the Englishman Count Rumford published his comprehensive essay, "Chimney Fireplaces." His main contribution was the alleviation of the smoking chimney. One fault in chimney design, he correctly asserted, was due to too large a chimney throat. Rumford also introduced the inclined fireback, which increased fireplace efficiency by providing an area of greater radiation. For the purpose of breaking up the current of smoke in the event of chimney down draft, the back smoke shelf of Rumford's improved fireplace ended abruptly— a practice strictly adhered to by fireplace masons to this day.

It is more important for the owner-builder to understand the aerodynamics of combustion and ventilation than it is for him to be presented with detailed specifications for one particular fireplace that works. In actual practice fireplace operation involves: (1) the motion of air toward the fire, (2) its passage through and over the fuel bed, (3) the mixture of ventilating currents with combustion products, and (4) the flow of chimney gases up the flue.

The first operational consideration involves the motion of air (*the draft*) toward the fire. Tests carried out by the Domestic Fuels and Appliances Committee, in England, indicate that the required volume of room air for average fireplace draft consumption is about 3,000 cubic feet per hour, which amounts to about four complete air changes per hour in an average-size living room. This same research-study agency found that the amount of fresh air required for ventilation by a family of four is about 2,400 cubic feet per hour. In other words, the operation of a standard fireplace will affect the displacement of over twice the amount of room air required for optimum ventilation. Half the amount of room air should, therefore, be drawn from the outside, be permitted to pass only indirectly through the building, and be prevented from immediate escape through the fireplace. Fireplace installation in our modern, tightly constructed, efficiently weather-stripped houses creates a problem of the availability of sufficient air for chimney draft. A partial air vacuum in these close constructions is the result, tending to pull smoke and combustion gases back into the room.

Air-intake control is the key to efficient fireplace combustion. The ignition of a correctly proportioned gas-air mixture will promote the complete combustion of the wood and emit "clean" gases containing only noncombustible carbon dioxide, water vapor, oxygen, and nitrogen.

Dry firewood has a high oxygen content, requiring a small amount of air for combustion. This explains why wet (green) wood has less heating efficiency than well-dried wood has. The point of combustion for wet wood can be lowered only by increasing the heat-appropriating draft. Freshly felled wood has a high moisture content, containing 50 percent water, and this wetness interferes with combustion. Evaporating water forms around the wood like a sheath of vapor and blocks the entry of oxygen to the fire. This results in a lowering of both the ignition and the combustion rates. If you must burn wet wood, you will need quantities of draft. For dry wood burning, however, the draft must be controlled.

When smoke and soot are observed coming out of a chimney, one can be certain that combustion is incomplete. What actually is seen are small quantities of hydrocarbons and free carbon (soot), which are not burned. As a result, much of the heating capacity of the wood fuel is lost. The heat loss is twofold: in hot gases which escape up the flue and in the unburned combustible particles. The first principle of fireplace design, then, is to aim at complete combustion.

The problem of foot-chilling floor draft was recognized and was, consequently, solved by Savot through his use of a subfloor inlet return for previously heated air. If this application were made today, this technique would eliminate the trouble frequently encountered in modern homes where fireplaces operate at low efficiency and have a tendency to smoke because of a deficiency in chimney draft. Cold air currents within a room could be minimized by a fireplace that would derive its draft independent of the air changes within the room. Combustion efficiency would increase with the incoming draft, consisting of preheated air. Without being aware of Savot's work, Wendell Thomas, in our day, has placed a return-air inlet in the floor by his heater stove, as part of his no-draft floor heating system, noted in chapter seven.

The owner-builder, desirous of eliminating foot-chilling floor drafts, should certainly consider the many advantages of a sunken hearth. The 12-inch-high raised hearth, commonly built nowadays, puts the heat radiation level that much higher off the floor, which remains cold. A sunken hearth also makes a welcome seating alcove, a higher flue is achieved and there is less danger from flying sparks, since the trajectory of a spark from a sunken hearth is less than the course of one from a raised hearth.

The passage of air through and over the fuel bed is promoted by use of a properly designed grate. A grate is employed to raise the firebed a few inches above the hearth; to feed air through the fire. It should be relatively small in size with closely spaced members to contain the wood charcoal formed during combustion. Charcoal must entirely cover the grate for complete combustion. If an over-sized grate is used, the tendency will be to build too large a fire, wasting gases which would, otherwise, be burned. Too small a grate would cause, on the other hand, problems with insufficient draft and create incomplete combustion, adding to the risk of incendiary tars being deposited in the chimney flue.

Radiation is more effective from a shallow fuel bed. In a deep fuel bed radiation is largely upward. Rumford found that, by sloping the fireback, much of the upward heat radiation could be redirected downward into the room where it is needed. Heat, radiating back across the firebed, further kindled the fuel on the grate by producing increased temperatures which considerably aided combustion and decreased smoke emission. To maintain high firing temperatures and a corresponding high efficiency, it is necessary to adequately insulate the fireback and the walls of the fireplace.

The admixture of ventilating currents with combustion products is one other aerodynamic consideration in fireplace operation. Count Rumford was the first to extensively study inside-fireplace proportions. His final rule of thumb was that the back of the firespace should be equal to the depth of the recess. Deep-set fuel beds produce more smoke than do shallow beds, since there is scant combustible air at the back of such a recessed facility. Experience has proven that the rate of smoke emission increases proportionally to the depth of the firebox, especially in the early stages of firing.

Count Rumford's conclusions about the relationship between the chimney throat and effective draft have yet to be scientifically questioned. The throat opening should be sufficient in size to constrain the effluent, so that it will be forced to pass at a speed high enough to discourage down drafts. Chimney-throat standards for most fireplaces require an 8-inch opening. Rumford recommends 4 inches. If a damper is installed in the larger-sized chimney throat, the situation is comparable. However, a properly designed construction makes a damper unnecessary. Simply stated, if the chimney throat is too large, some superfluous cool air, not affecting ignition or combustion, will be drawn over the fire to increase the smoke emission. Putnam, writing *The Open Fireplace in All Ages*, in 1886, recognized the truth in this relationship between effective draft and chimney-throat size:

Cold air, being heavier than warm, will fall below the latter and press it upward to make room for itself. Thus, air in the neighborhood of the fireplace will press the hot smoke up into the chimney throat. If this throat is only large enough to take the smoke, hot air only will enter the flue, and the draught will be rapid. But if the throat is larger than necessary, that part of the cool air of the room which enters the fireplace and becomes most heated by the fire, and next in buoyancy to the smoke, will, in turn, be pressed up by the cooler air behind it and enter the flue alongside of the smoke. Indeed, the entire volume of the air of the room, being warmer than the outside air, will tend to enter the flue with the smoke, so long as there be room provided for its entrance.

The English scientist Dr. P. O. Rosin has done much significant research on the aerodynamics of open fires. He built scale models of fireplaces, using celluloid sheets to reveal visually the behavior of the gaseous flow associated with open fires. He was the first to prove that Rumford's horizontal smoke shelf at the top of the fireback was aerodynamically faulty, producing eddies of smoke-laden air which back up into a room with the least down draft. Rosin found, also, that the

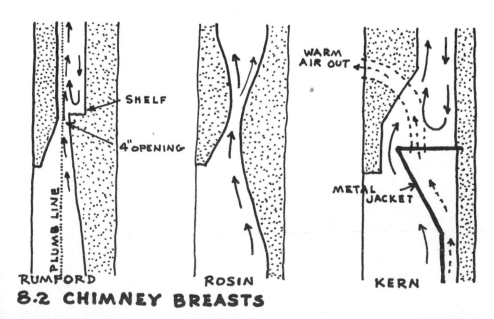

8.2 CHIMNEY BREASTS

chimney breast must be located and shaped in relation to the upper part of the fireback. A passage must be provided between chimney breast and fireback, which narrows at the throat and, then, gradually converges with the base of the flue. The lower edge of the chimney breast should be well rounded and free from abrupt changes of direction.

Some eddies will occur where the smoke shelf is minimized, as seen in Rumford's design. Rosin is correct to say that eddies will not occur where the smoke shelf is eliminated entirely, but neither will they occur where the smoke shelf is ample enough to allow the down draft a free return trip up the chimney. Rosin's design includes a curved, free-flowing chimney breast and throat. The principle is good, but impossible for the owner-builder to fabricate in metal in his home workshop.

The final aspect of aerodynamic fireplace design concerns the flow of gases up the chimney flue. A chimney performs a dual function. It creates the draft and eliminates combustional residues. The chimney should be designed to deliver just enough air by way of adequate draft for complete combustion within the firebox. The National Bureau of Standards' tests indicate that the draft-producing ability of a carefully designed and built chimney is substantially improved over that of conventional chimneys when the flue is reduced in the constricted area above the chimney breast and the fireback, before it enters the flue. Oddly enough, flue temperature is the major factor determining flue size:

> The formula for calculating draft is based upon "mean temperature" which is the average of the temperature at the base of the chimney and the temperature at the top. If the chimney construction is such that an appreciable amount of heat in the flue gases is lost through the chimney walls, the temperature at the base will have to be higher to produce a given mean temperature than if the chimney walls were to be built to resist the transfer of heat through them.
>
> If the walls of the chimney are massive and absorb the heat of the flue gases, a long time may be required to raise the average flue gas temperature high enough to produce the necessary draft. On the other hand, a long time is required to dissipate the heat

retained in the chimney walls and the draft will continue at high intensity even though no heated flue gases are supplied by the heating system.

The ideal chimney has no heat loss through its wall and, consequently, no difference in temperature of flue gases between bottom and top; no heat capacity or heat-retaining ability with, consequently, no time lag in producing draft when combustion is started or in "killing" draft when combustion ceases.

Obviously, the fireplace should be located on an inside wall of the building to be heated. The heat loss from placement of the fireplace on an outside wall is something like 25 percent. A chimney that is exposed to the weather along its entire length on one or more sides is bound to cool off when the fire is low. Then, when the fire is kindled again, the products of combustion have to try to force their way out of a chimney filled with dense, cooled gases. Moreover, the heat that escapes from an inside chimney will help to warm the house.

The National Bureau of Standards has demonstrated that a round 7-inch flue, with a cross-sectional area of only one-half that of a rectangular 9- by 12-inch flue, will just as readily produce the necessary draft, since corners in the larger square or rectangular flue sections have little effect on draft. This means, of course, a substantial saving for the builder in flue tiles, surrounding masonry, and labor costs.

Often installation of some form of draft diverter becomes necessary to prevent down-draft tendencies at the flue terminal. A steady draft is necessary if proper combustion is to be assured over long periods of firing. One type of draft diverter consisting of a permanent annular opening is especially effective under either normal or down-draft conditions. An even simpler way to counter down drafts is to reduce the flue terminal, that is, the chimney top. The velocity of flue gases is, thereby, increased at the flue terminal and is sufficient to oppose tendencies to down draft. Tests show that insulation between the flue lining and the surrounding masonry improves the chimney draft.

Fireplace construction today is something of a paradox. The growing use of heat-circulating forms at the expense of progress in the construction of classic fireplace design is of some concern. As efficient as these commercial forms are, they drastically restrict originality in

8.3 FIREPLACE PROPORTIONS

No.	A	B	C	D	E	TO 13' F	TO 13' G	TO 26' F	TO 26' G	TO 39' F	TO 39' G	SQ. FT.	CU. FT.
1	24	20	13	14	4	10 × 10		8 × 8		8 × 8		70-240	1400-2100
2	30	24	15	19	4	10 × 10		8 × 8		8 × 8		240-320	2100-3200
3	36	27	16	24	5	15 × 10		10 × 10		10 × 8		320-430	3200-4200
4	42	30	17	29	5	15 × 15		15 × 10		10 × 10		430-540	4200-6400
5	48	33	19	33	6	15 × 15		15 × 10		10 × 10		540-750	6400-8800
6	52	35	20	36	6	15 × 15		15 × 15		15 × 10		750-	8800-

FLUE SIZE FOR STACK HEIGHTS — ROOM SIZE

design and the resulting character of the fireplace. They do, however, dramatically increase room heat. A number of commercial units are available. Heatform, Heat-a-lator, and Majestic are the more popular brands. The heat-circulating form consists of a steel fireplace shell surrounded by an outer jacket of steel. The double walls, so formed, trap cool air against the hot firebox wall, where it is heated and rises through grilles into the living space. Cool air is drawn from the same room through other grilles in the floor, where it continues toward its eventual heating in the double-walled air jacket of the heat-circulating form. A continuous, circulating stream is, thus, formed to heat and reheat room air. An even more effective warm-air distribution can be achieved by placing a circulating fan in any one of the ducts to push the entire action forward.

These heat forms, today, cost from $150 up. Therefore, one enterprising owner-builder, whom I know, fabricated his own heat-circulating form, using raw materials from a junk yard. It performs effectively and probably delivers as much usable heat as the most efficient and expensive commercial unit on the market. Following is a detailed description of this home-fabricated heat-form construction, along with a sketch showing some design variations for possible exterior treatment:

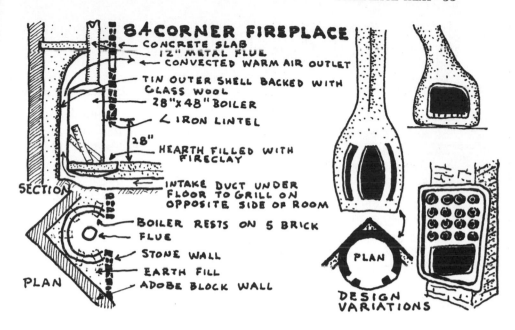

84 CORNER FIREPLACE
- CONCRETE SLAB
- 12" METAL FLUE
- CONVECTED WARM AIR OUTLET
- TIN OUTER SHELL BACKED WITH GLASS WOOL
- 28"x48" BOILER
- ∠ IRON LINTEL
- HEARTH FILLED WITH FIRECLAY
- INTAKE DUCT UNDER FLOOR TO GRILL ON OPPOSITE SIDE OF ROOM
- BOILER RESTS ON 5 BRICK
- FLUE
- STONE WALL
- EARTH FILL
- ADOBE BLOCK WALL

SECTION

PLAN

28"

PLAN

DESIGN VARIATIONS

The principal ingredients of this construction were a discarded water boiler, about 24 inches in diameter by 4 feet high, plus several discarded range boilers, which were 12 inches in diameter. These can be found in abundance in most junk yards. The larger boiler must be cut down as shown and have the upper part closed off with a ⅛-inch-thick metal plate.

The combination boiler-firebox was set in the corner of an adobe-wall room, in the manner seen in Mexican homes. A 12-inch-round metal flue was welded in place atop the metal-capped boiler. Prior to this aspect of the construction, a 12-inch air duct was formed under the floor from the far side of the room, culminating in a distribution plenum which was slightly larger in diameter than the firebox above it. The firebox was supported over the plenum by spaced bricks, which allowed free air circulation to the heating space behind the firebox walls. An outer jacket of sheet metal was placed behind the firebox, making a sealed, 2-inch airspace. It was backed by a layer of salvaged glass wool and earth.

Stone jambs were carried up from floor level, adjacent to the vertical edges of the firebox. At lintel height, steel angle iron was installed for support of the masonry above. A solid face wall of stone was carried upward above the lintel, pierced only by the 10- by 16-inch warm-air outlet. Two feet above the top of the firebox, a small concrete slab was cast in place, resting in grooves cut in the outer walls and in the stone face wall. This slab locked into position the fireplace flue and capped the air-heating chamber. Above this slab, the triangular space formed between outside walls and the face wall, divided only by the ascending flue, was filled with earth. Where the flue exited through the roof, a mesh-reinforced collar of concrete, 5 inches thick, was formed around it.

The circular hearth, made by the lip remaining at the lower part of the original boiler, was then poured with a mix of fire clay and Portland cement and was allowed to set. Wood was placed in the upright firebox in a vertical leaning position, in the manner typical of Mexican fireplaces. Fairly long logs were so placed, and, though some of the burning took place in the head of the firebox, out of sight, the heat produced was utilized to warm jacketed air, which radiated into the room with considerable warmth through the hot-air outlet.

Sheet iron is 12 times more conductive than stone masonry. A metal-jacketed heat chamber will, therefore, emit quantities of conductive heat which would otherwise be lost through absorption into the masonry. A simple, easy-to-build fireplace form can be built in the home workshop for about $15 worth of materials. The unit illustrated below is the result of improvements I have made in over a dozen years of work. It represents the best in fireplace research and thinking and is designed to give maximum heating efficiency.

The metal chamber may be cut, bent, and welded out of a single 3- by 9-foot piece of sheet iron. A damper, welded to a pair of hinges, is, in turn, welded to the smoke shelf. A cool-air-supply duct should be provided at the front lower sides or at the rear of the lower back of the unit. The warm-air outlet is best placed at the upper front of the fireplace. Outlet ducts, built into the floor slab or into the attic space, can be positioned in an adjacent room when a forced-air circulation fan is employed to distribute the hot air through controlled outlet grilles.

If the fireplace is not to be the only source of heat, an exceedingly effective, economical heating arrangement is the incorporation of the forced-air heating system with the fireplace system. In this instance,

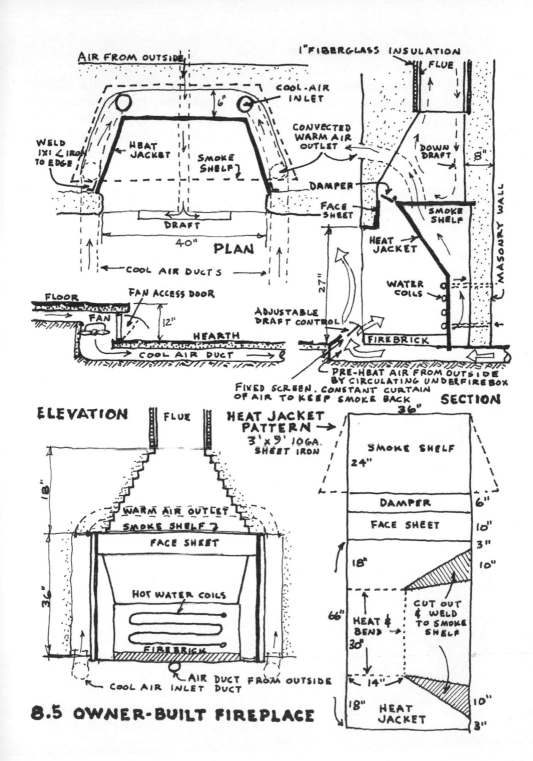

AIR FROM OUTSIDE

COOL·AIR INLET

6"

CONVECTED WARM AIR OUTLET

1"FIBERGLASS INSULATION
FLUE

DOWN DRAFT

8"

WELD 1X1 ∠IRON TO EDGE

HEAT JACKET

SMOKE SHELF

DAMPER

FACE SHEET

SMOKE SHELF

MASONRY WALL

DRAFT

40"   PLAN

HEAT JACKET

COOL AIR DUCTS

WATER COILS

27"

FLOOR   FAN ACCESS DOOR

FAN   12"

ADJUSTABLE DRAFT CONTROL

HEARTH

COOL AIR DUCT

FIREBRICK

PRE-HEAT AIR FROM OUTSIDE BY CIRCULATING UNDER FIREBOX

FIXED SCREEN. CONSTANT CURTAIN OF AIR TO KEEP SMOKE BACK

SECTION

ELEVATION   FLUE

HEAT JACKET PATTERN →
3' x 9' 10GA. SHEET IRON

36"

18"

WARM AIR OUTLET

SMOKE SHELF

FACE SHEET

36"

HOT WATER COILS

FIREBRICK

AIR DUCT FROM OUTSIDE

COOL AIR INLET DUCT

SMOKE SHELF   24"

DAMPER   6"

FACE SHEET   10"

3"

18"   10"

66"   CUT OUT & WELD TO SMOKE SHELF

HEAT & BEND →

30"

14"

18"   HEAT JACKET   10"

3"

8.5 OWNER-BUILT FIREPLACE

the fireplace hot-air jacket chamber acts as a plenum, with furnace heat distributed through the fireplace heat form. A separate fan control, a summer switch, is installed, so that the furnace blower can be used to distribute fireplace heat without the use of the furnace itself. Each can be used separately, or both heating units can operate simultaneously.

The floor plan illustrated below will be referred to throughout this book. It is representative of the best thinking to date on quite a number

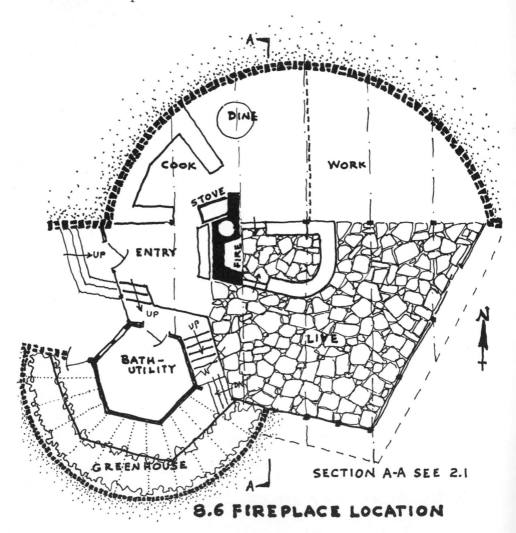

SECTION A-A SEE 2.1

8.6 FIREPLACE LOCATION

of owner-built home features. Cross-section A-A, drawing 2.1, has been taken, in part, from this plan. The reader should note, first, the central location of the masonry core. A conversation pit has been drawn around three sides of the fireplace opening. The cook stove, illustrated in chapter six, is located on one side of the masonry core. Water coils, in both the fireplace and the stove, circulate hot water into a storage tank located between the two units. Flues from both the stove and the fireplace are positioned in the single masonry core. Lastly, air-intake slots around the perimeters of the rooms return cool air under the floor toward the fireplace pit. A series of silent, squirrel-cage fans direct the air either through the fireplace heat jacket or through the cookstove-heater jacket to be warmed and recirculated again through the entire system.

If an owner-builder really expects to take his fireplace heating system seriously and if he desires more than "atmosphere" from his wood-heat supply, then he should conscientiously consider these design features. The stove-and fireplace system, here discussed, is designed to keep this prototype house comfortable in the most inclement weather, when it is combined with the sun-tempering and earth-berm-insulating features, which were covered in previous chapters.

9 LANDSCAPE DESIGN

Hopefully, the current liberation of the sexes in our social, economic, and political arenas will influence domestic architecture as well. If one agrees with the premise outlined in the previous chapter that interior space speaks of femininity while outside space speaks of masculinity, then the new design concept that interior space should be reciprocally and harmoniously extended to and connected with outside space may, at the same time, be indicative of this contemporary era of liberation. Knowledgeable landscape architects claim that the very same principles of design apply to outside planting design. Every plant, no matter what form it may have, is both a construction in space and an enclosure of space.

As an enclosure of space, plant forms extend the walls, floors, and ceilings of a room to become outdoor hedge walls, lawn floors, and tree ceilings. Outdoor shelter forms, such as arbors, pergolas, and pavilions, find shelter-counterparts within the house. As constructions in space, the sculptural effects of foliage, rocks, and pools can be likened to the sculptural quality of the house architecture and to the decorative furnishings present in the building's interior.

This concept of integral building and planting was actually practiced by eighteenth-century Chinese. Called "Feng shui," the basic principle was derived from the teachings of Lao-tse, the sixth-century Chinese philosopher, who taught a return to nature. Nature and humanity were harmonized in the Chinese garden, the house serving mankind's practical and serious needs, and the garden being a place for freedom of expression and spirit. In the house, persons are in the society of fellow beings, the Chinese thought, but, in a garden, one is in the society of natural forms.

It has been said that inside the house, a Chinese is a Confucian, adhering strictly to the conventions and to the moral codes set down by Confucius. In the garden, one is a Taoist, exemplifying the primitivistic, libertarian precepts of Lao-tse. It is interesting to note that, while the Chinese home is orderly and formal in style, perhaps limiting to the spirit, Chinese garden forms are irregular and sinuous, inspiring the spirit's release. According to Wing-tai Chan, the Chinese garden is a place "where man laughs, sings, picks flowers, chases butterflies and pets birds, makes love with maidens, and plays with children. Here, he spontaneously reveals his nature, the base and the noble. Here also he buries his sorrows and difficulties and cherishes his ideals and hopes. It is in the garden that men discover themselves. Indeed, one discovers not only his real self but also his ideal self—he returns to his youth. Inevitably, the garden is made the scene of man's merriment, escapades, romantic abandonment, spiritual awakening or the perfection of his finer self." In Western gardens, we seek the comforts and conveniences which people have come to consider essential to their entire experience of well-being, for, after all, it is the activity of people that determines the form and the character of garden planting and sculpting.

Modern landscape designers employ many devices to satisfy contemporary tastes. To summon garden beauty we arouse interest by planting in variety. To excite we plant in sequence. To stimulate we plant in color. A shrub can be planted to create many effects, depending upon its placement and its relation to human scale. If the plant is above eye level, it can function as protective enclosure. If it is kept at chest height, the effect is more one of spatial division. If the planting is merely waist high, it functions as traffic control. If knee height, it gives a directional aspect to the planting. It is the human scale, in this case, a person's height, that measures and relates the garden elements, including walls, fences, trees, and shrubs. The human line of vision analyzes whether these landscape elements provide privacy, separation, or direction.

Eckbo is surely the most noted representative of the modern landscape movement. His book, *Landscape and Living,* is a clear statement and a concise presentation of modern landscape objectives and practices. Eckbo gardens are beautiful designs of plant-structure relationships, and they contain all the amenities so eagerly sought by present-day home owners. In all of his gardens, one will find that the plant and the structural elements are well selected. The groupings, plant forms, and their massing are well arranged. The whole scheme is practical to maintain.

Outdoor living with minimum maintenance and maximum charm, however, gives little impulse for people to seek spiritual uplift from that essential, revitalizing contact with plant growth and with the fecundity of the earth. Gardens in the Orient capture this essence,

9.1 RUDOFSKY WALLS

sojourners there gain strength and inspiration. One, also, finds few modern garden designers with any concept of *Spieltrieb*, the German term relating to the playful instincts expressed in plant forms, garden structures, and organization. The idea that a garden can be the site of gaiety, of imagination, of fantasy, as well as being a place of meditation and repose seems alien to modern thought and practice.

However, great respect should be given to one architect, who successfully expressed the *Spieltrieb* concept in a garden plan for a modern Italian muralist client. Bernard Rudofsky speaks of his design in these terms:

> A free-standing wall, plain and simple, with no special task assigned, today is unheard of. In a garden, such a wall assumes the character of sculpture. Moreover, if it is of the utmost precision and of a brilliant whiteness, it clashes—as it should—with the natural forms of the vegetation, and engenders a gratuitous and continuously changing spectacle of shadows and reflections. And aside from serving as the protection screen for the surrounding plants, the wall creates a sense of order. Three abstract murals compete with the umbrageous phantasmagories.
>
> . . .
>
> An old apple tree pierces one of the walls, lending it (methinks) a peculiar monumental quality. The pergola is reduced to almost linear design, and does not intend to more than assist and coordinate. A wisteria has taken possession of it in the space of a few months; bamboo shades are hung from it in summer. The wiry appearance of the poles is accentuated by bright colors.

Another exceptional landscape architect, Roberto Burle Marx, expresses the *Spieltrieb* concept in bold and positive terms. His designs are curving, free-form reactions opposing symmetry and rectangularity. One of the more interesting things about Burle Marx's gardens is the attractive use of native plants, plants considered to be mere weeds among other gardeners. He searches his native, Brazilian jungles for indigenous plants and combines their placement with skillful use of stone mosaics and waterpools.

The central purpose of this chapter is to offer the home builder a working outline for landscaping his new home. For many years, data has been collected which can be used as a basis for good planting-design procedure. This approach has not been along modernistic landscaping lines, nor has it tried to analyze the subjective and symbolic forms of traditional gardens of the Orient. Rather, the purpose has been to organize a planting-design procedure which is based entirely on the ecological use of natural vegetation. The emphasis should be on the relationship between plants, climate, and soil, as well as between one type of plant and another. Once this harmony is created, garden beauty and spiritual release are naturally forthcoming. Every experience, from frolicking to quiet repose, may then ensue from spontaneous expression of the personal preferences of the garden's inhabitants.

Rudolf Geiger is one of the earliest climatologists to indicate the direction of this new concept of planting design. His excellent study of the microclimate of the site indicates the procedures and the methods necessary for achieving this new garden form. He found that a mixed forest growth of spruce, poplar, and oak shades the ground from 70 percent of the sun's heat. Forests, in summer, are cooler than cleared land and warmer than cleared land in winter. Nature protectively covers the earth's soil with vegetation. Heat, otherwise held by the soil to the detriment of fragile soil microorganisms, is dispersed by the screen of heat-deflecting foliage. Heat is dispersed from the soil, incorrectly presumed inanimate, by the thermal deflection of ground-covering plants and mulch, creating an atmosphere of summer cooling and one of protection against winter cold. By deflecting cold winds from surfaces, an evergreen windbreak effectively prevents heat loss from buildings and garden areas. Drifting snow may be controlled by strategically planted evergreen hedges.

Barren housing-tract developments leave one to wonder how such a basic climate-control device as that of trees used for summer shading effect can be ignored by so many builders. But even a thorough, intellectual understanding of how deciduous trees provide generous shade at the appropriate summer season and of how they, fittingly, lose their leaves in the autumn so that the sun can easily penetrate leafless branches throughout the winter is not enough to assist the amateur home builder in his building construction and his selection and placement of trees. Climate-control experts employ a heliodon to determine the most desirable, positive location for structures and for the vegetation that is to be planted around those structures. The Olgyay brothers, at one time professors of architecture at Princeton University, published more vital information on climate control than any other combination of research efforts. In this regard, one is encouraged to see, specifically, their book, *Solar Control and Shading Devices.*

The owner-builder who prefers to approach his landscape planting soberly, in a creatively calculated manner, should be alerted to the use of a model design aid which will indicate the amount of light penetration and the extent of shading for nearly every phase of his garden and house planning—at his particular latitude, for any day of the year and for any hour of the day. This simple-to-build sun machine,

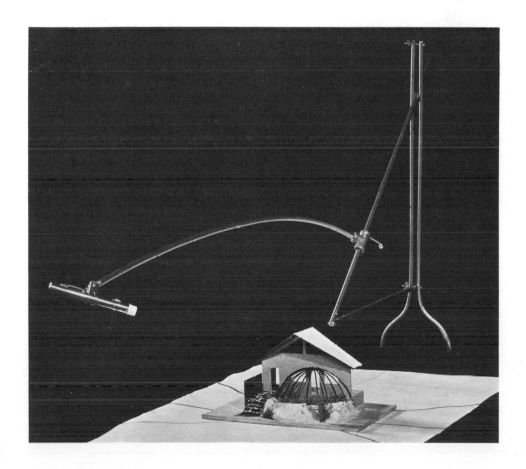

9.2 HOME-MADE HELIODON FROM 5/16" CURTAIN ROD

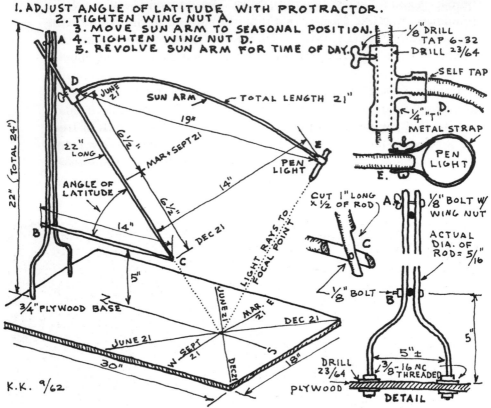

1. ADJUST ANGLE OF LATITUDE WITH PROTRACTOR.
2. TIGHTEN WING NUT A.
3. MOVE SUN ARM TO SEASONAL POSITION.
4. TIGHTEN WING NUT D.
5. REVOLVE SUN ARM FOR TIME OF DAY.

K.K. 9/62

the heliodon, can accomplish this impressive task at little cost in materials.

A heliodon is a simulated-sun device. When using it, one must make a cardboard, scale model of the proposed garden or house in plan form, that is, at first omitting the outside walls and roofs of any structures. Then, as various summer or winter solar angles are set on the heliodon, the designer's addition of walls and roofs and specific plantings will cast shadows on one's model, determining window sizes, ceiling heights, plant groupings, and plant spacings for eventual construction or placement.

Soon after beginning to work with a heliodon, one discovers that east and west building openings or garden aspects are vulnerable to summer solar-heat radiation. In late June, the sun is at its zenith and the day is at its longest length of time. The horizontal sweep of the summer sun is 242°, about twice that of its sweep in the winter. One might readily realize from model and heliodon manipulation that certain east and west openings and aspects will be impossible to protect from summer-sun heat by conventional means. Illogically, something like a 20-foot roof overhang might be indicated for shading an exposed area! It will, therefore, instantly and graphically become obvious that the judicious planting of a few particular kinds of trees and shrubs in combination with the building of complementary, external shading devices or garden structures will achieve the desired results.

If special solar orientation problems occur, it may be necessary to shift the house site or to relocate vegetation and shading devices on this experimental model. One of the attractive features of house designing using model and heliodon is that this use introduces an element of play while one visually and actively manipulates the various

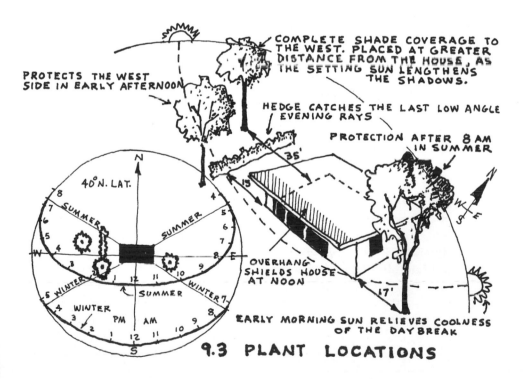

PROTECTS THE WEST
SIDE IN EARLY AFTERNOON

COMPLETE SHADE COVERAGE TO
THE WEST. PLACED AT GREATER
DISTANCE FROM THE HOUSE, AS
THE SETTING SUN LENGTHENS
THE SHADOWS.

HEDGE CATCHES THE LAST LOW ANGLE
EVENING RAYS

PROTECTION AFTER 8 AM
IN SUMMER

40°N. LAT.

OVERHANG
SHIELDS HOUSE
AT NOON

EARLY MORNING SUN RELIEVES COOLNESS
OF THE DAYBREAK

9.3 PLANT LOCATIONS

elements of one's dream. One lives with his three-dimensional creation as separate spaces and separate effects, in juxtaposition, eventually join in a completed whole—the physical representation of one's intentions.

The shape and character of the shade tree will determine the extent and shape of its shadow. The variety of tree chosen will, therefore, depend upon the shape of the area to be shaded. The maple and the ash produce circular shadows, in summer, with an ascending branch pattern in winter. Honey locust and tulip trees have oblong shapes. The white oak is wide and horizontally oblong, with an open-branched structure. The Lombardy poplar is columnar, in appearance and the American elm is vase-shaped. Other shade-producing trees are the weeping willow, Russian olive, flowering dogwood, sweet gum, American beech, maple, white birch, and the Siberian crab apple.

Landscape gardeners seldom recognize the effect that plants have on the heat and moisture content of the soil and the surrounding atmosphere. As a result, one mistake that is usually made involves the planting of shrubbery too close to the house. The density of the shrubs prevents breeze penetration, reducing evaporative cooling and causing high temperature and high humidity to persist within the foliage. Conversely, trees and grass planted in relative proximity to the house will, during seasonal heat, allow heavier, cool air to flow inside through accommodatingly designed window openings. Leaves and grass naturally absorb solar radiation, and the resulting evaporation cools surrounding air. Mowed turf is a good climate-control planting. By its shading of the soil, heat absorption is prevented and intensive reradiation is eliminated.

Dr. Robert Deering, University of California professor of Agriculture, reports that when trees are planted near the south glass wall of a building, several desirable effects occur. The north side of the tree facing the south wall of a building, is the chilling side of the tree. Its effect is to cool the inside of the house. Annoying glare is also substantially reduced by a similar placement of trees. Air-borne sounds can be effectively reduced by densely planted trees and shrubs, and the viscous surface of leaves catch dust to function as air filters.

In Europe, more than in this country, vines are grown to provide seasonal shade and to control evaporation. Vines are particularly effective when grown against west walls or on trellises near the west wall of a building. Some interesting and attractive vines are clematis, bittersweet, frost grape, parthenocissus, hydrangea petiolaris, wisteria, silver lace vine, Chinese fleece vine, Dutchman's pipe, forsythia, and ipomoea.

Color-fragrance relationships in planting design are a new consideration, and, of particular note here, is the psychological effect created by the dark, dense, glossy greens which are prominently used in areas of high humidity. There is a tendency to accentuate an oppressive climate through duplication of the effect of the atmosphere by selecting correspondingly heavy plants. This effect should, however, be alleviated by the inclusion of lighter, rarer colors of greenery. Likewise, thinner plant shapes, which permit the incursion of light and warmth, should be encouraged in cool climes. In hot, dry zones of low humidity, natural vegetation is sparse, dull, and fuzzy. Gray, gray-green, and brown-green colors predominate. In this climate, it is enlivening to entice a spark of bright, clear, green growth. Large, thick-leaved foliage also will feel cool and moist in the encircling atmosphere and to the touch—a most refreshing experience in arid regions.

An enlightened approach to planting design requires a thorough understanding of one's region and site. This basic understanding, which includes information about weather patterns, soil conditions, and native plant life, must necessarily precede an intelligent and a responsive treatment of landscaped environs. The primary objective of planting design is the creation of an aesthetically pleasing and a climatically comfortable environment.

10 THE PLAN

When future archaeologists study the remains of our culture, most of their data will be amassed from the unearthing of our buildings and our dwellings and their contents. The level of our aesthetic achievement will be manifest in the design of our buildings and in their decoration. The degree of our technical development will be evident in the engineering of our constructions and in our application of materials. The societal functions of family life and community organization will be learned from a study of the plans of our private homes and public quarters.

Today, we exhibit numerous house styles: Cape Cod, Southern Colonial, Pennsylvania Dutch, Greek Revival, English Half-Timbered, French Provincial, Spanish Monterey, and California Ranch Style. The floors of buildings vary in array from single-story to split-level to two-story levels. Students learn to functionally classify houses in architectural terms such as Eclectic, Expressionistic, Rationalistic, Structural, Brutalistic, Functionalistic, and Organic. Yet, when these various buildings are viewed in plan, they cannot be differentiated.

The present-day house plan is said to reflect a casual, informal living pattern, its functional aspects being limited, primarily, to reproduction, nutrition, and survival. Facilities which were fundamental to the organization of nineteenth-century homes for the satisfaction of the family's economic, religious, educational, and social needs are, by contemporary arrangement, now found unsuitable for the majority of families. Increased mobility, primarily due to the development and use of the automobile, has today made it possible, if not desirable, for family members to go beyond the home to seek satisfaction of these needs within more elaborate and specialized facilities. In addition, while the automobile contributes to the alienation of the family from its homeplace, it tends to alienate them also from community life in their residential neighborhood. It is not unusual for the suburban father to work in an urban area, for the mother to shop and spend leisure time in another area, while the family goes to church and the children are educated in yet other areas. In our industrialized society, lengthening of the life span and the decrease in family size are several other factors that distinguish present living patterns and today's architecture from those of yesteryear's plow culture. Architecture in the cities and suburbs, consequently, subscribes to unimaginative solutions and to blatant decorations and embellishments which are the products of confined space: namely, of the box. This symbol of current living patterns is ideally suited to our notion of the efficacy of packaged goods, personal isolation, and compartmentalized thinking and residency.

The box house, containing smaller, internal box rooms, is, of course, not new in mankind's scheme of things. In Germany, the man who builds has, for centuries, been called a *Zimmerman,* a room man or carpenter. The contemporary builder of tract homes is such a room builder, and, as such, he is qualified to satisfy conventional tastes and space requirements. His experience, efficiency, equipment, and general knowledge of room construction can hardly be surpassed by even the most earnest owner-builder.

An increasing number of architects, however, maintain that space should flow, that it should not be parceled into boxlike cubicles. An advanced designer, like Leonardo Ricci writing in *Anonymous (20th Century)*, will contend that our architecture eventually will change from its boxlike quality to that of freely flowing space. It will change involuntarily as a consequence of changes which will take place in peoples' living patterns. Critically needed for the inception of a new architecture are new structural concepts and new space proposals. Lao-tse was, perhaps, the first to counsel men that their needs will become elementary once they achieve a freedom of space, which will be acquired not by acquiescence to the demands of positive, space-enclosing components such as walls, floors, and roofs, but by attention to negative qualities intrinsic to form, movement, volume, and—emptiness. This concept is embodied for us in these words from Lao-tse:

> Thirty spokes are made *one* by holes in a hub,
> by vacancies joining them for a wheel's use.
>
> The use of clay in molding pitchers
> comes from the hollow of its absence.
>
> Doors and windows in a house
> are used for their emptiness.
>
> Thus we are helped by what is not
> to use what is.

We may deduce from the expression of this ancient luminary that the role of the professional dwelling builder of today would be considered that of a dead-space decorator. Artificial solutions are too often invented as substitutes for the concrete solutions urgently required by fundamental problems in architectural design. Knowledge of the concept of freedom of space can, however, enable one to design natural and effective solutions into his housing requirements.

Competent builder-designers struggle for freedom of space with the same determination that political revolutionaries battle for freedom from oppressive government. The stifling system of the six-planed enclosure (the four-walled room), the tyranny of the T-square, the ticky-tacky decoration of the box house can be as personally enervating and demoralizing as life under despotic rule.

A fundamental understanding of the characteristics and the qualities of space will enable us to organize that space and to condition it through building design and construction that is geared to its wise use and its ultimate enjoyment. By adding some modern concepts to the ancient system of space analysis devised by early Chinese, one is able to comprehend the unified whole, the *Gestalt*. The schematic form presented below can be utilized in the inception of one's own building plan. Chinese found life more meaningful when they related their houses as well as their cities to the time of day, to the seasons, and to solar orientation. They found meaning in subtle adjustments of the positioning of open-enclosed and active-passive spaces.

Indirectly, these hourly, seasonal, sun-oriented relationships are a subconscious part of man's experience, in the same manner that the need for privacy and for sociability are, by turns, part of our subliminal experience. A glass house, for instance, with its openness and translucency would be conducive to satisfaction of the socially extroverted aspect of man's nature, of his longing for expansiveness. Man's introverted nature might, on the other hand, also seek the confines of his dark and mysterious cave origins. Somehow, the enclosedness of one space must not conflict with the expansiveness of the other. Both are equally essential to optimum living satisfaction and appreciation.

Space does not, necessarily, have to be confined by the six planes of the conventionally designed and built room. Space can be boundless, or it can be only partially enclosed. As schematically illustrated on the following page, open ectospace overlaps enclosed endospace to form a transitional, partly open, partly enclosed mesospace.

Found in all living things, the passive-active state must be another of our architectural considerations. Passive areas of our home environment allow expression of the introspective, introverted aspect of our lives and personalities, while our extroversion finds expression in activity areas that are both indoor and outdoor. Passive-active relationships are the essential rudiments justifying a functional division of space. They are followed in importance by a more involved sequence of space use. For the purpose of classification, these space activities can be listed by the degrees of public and private life that they offer. Gradations of these activities consist of the public, semipublic, operative, semiprivate, and private.

The first step to developing building plans requires schematic analysis by owner-builders of the space-use activities engaged in by their family. Related activities should be grouped into use areas. By zoning related activities, maximum livability is realized and each activity is carried on without interference from unrelated activities. In place of an actual wall division, each activity would receive an expandable

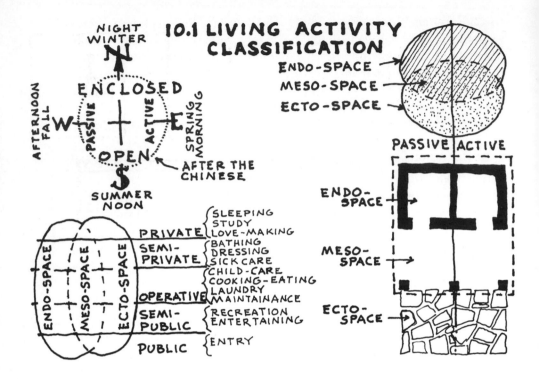

10.1 LIVING ACTIVITY CLASSIFICATION

space allocation. A series of movable room dividers and visual and auditory buffers dividing a space for a variety of activities will create areas for multiuse with an economy of means.

Contemporary house planners call the concept of flowing space "open planning," as distinguished from that space which is cut up into separate cubicles. The practical application of this theory effectively increases one's usable, interior space. Visual space and usable floor area are, essentially, enlarged with demarcations made by adjustable partitions rather than those made by solid, stationary walls. The corollary of open planning is its economy of space. This flexible, multiuse of space with its overlapping activities can, indeed, effectively reduce the overall need for floor space, commonly called square footage, and can, ultimately, reduce building size and cost.

Increased space needs necessitated by family enlargment and growing children can be met by an interior construction capable of expansion and contraction with those changing circumstances. Sociologists

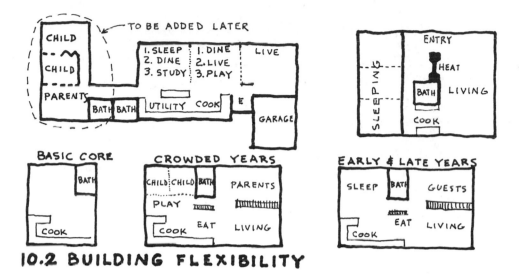

10.2 BUILDING FLEXIBILITY

delineate three family-life stages: the early, the crowded, and the late periods. Each stage represents a different requirement for building space, further illustrating how inadequate the room division of space really is. The space-squeezed family can, conceivably and at considerable extra expense, move to a larger house and, in turn, to a smaller place as the family size decreases, or, as a traditional solution, attic and basement areas can be renovated to accommodate additional space requirements. Another way of coping with this matter is for one to design the house for its estimated, maximum capacity, and, then, to rent out the unused additional space during the early and late years. A more economical solution for the young owner-builder is, however, to start one's building with a core for cooking, living, sleeping, and bathing. As the family size waxes and wanes, space can, first, be added to this core and later subdivided to accommodate activities occurring in later years.

It was Frank Lloyd Wright who first introduced important open-planning concepts into residential design. For a more spacious feeling, he integrated living, dining, and kitchen functions. The kitchen-work areas were centrally located between the living and sleeping areas. Wright's entries and hallways consisted of well-planned circulation patterns. From the main entry, one had immediate access to every

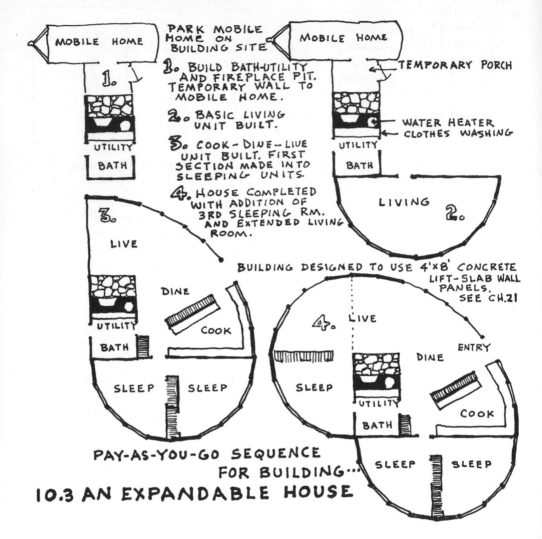

PARK MOBILE HOME ON BUILDING SITE

1. BUILD BATH-UTILITY AND FIREPLACE PIT. TEMPORARY WALL TO MOBILE HOME.

2. BASIC LIVING UNIT BUILT.

3. COOK-DINE-LIVE UNIT BUILT. FIRST SECTION MADE INTO SLEEPING UNITS.

4. HOUSE COMPLETED WITH ADDITION OF 3RD SLEEPING RM. AND EXTENDED LIVING ROOM.

BUILDING DESIGNED TO USE 4'x8' CONCRETE LIFT-SLAB WALL PANELS. SEE CH.21

MOBILE HOME

TEMPORARY PORCH

WATER HEATER CLOTHES WASHING

UTILITY

BATH

LIVING

UTILITY

BATH

LIVE

DINE

COOK

SLEEP   SLEEP

LIVE

SLEEP

DINE

ENTRY

UTILITY

BATH

COOK

SLEEP   SLEEP

PAY-AS-YOU-GO SEQUENCE FOR BUILDING...

10.3 AN EXPANDABLE HOUSE

activity area of the house. The hallway functioned as passageway, storage facility, and utility area. Wright was, perhaps, the first to build the house with an exposure differing from the traditional outward one. He preferred an orientation to private view.

Wright introduced into house design and structure many important planning concepts. Even 50 years after their inception, these ideas have barely trickled down into the mainstream of conventional architectural practice and expression. Wright spoke of "organic architecture" and of the "natural house." Today, architectural renegades surpass Wright's perceptions to somewhat the same extent that he surpassed his contemporaries. These mavericks now speak of "free-form" architecture and the "endless house," a term coined by the architect Kiesler. The arrangements and the dimensions of the endless house are determined by the various living activities of its inhabitants. Its free-flowing form derives from the fact that any section of the house can be closed off or opened into one, continuous space. Unlike a room, this space cannot upon first encounter be fully perceived, measured, and comprehended by our faculties. It has a charming indefiniteness of bounds and an elusiveness of volume, which lend it an unpredictable, natural quality unlike anything else found on the contemporary scene.

11 THE FREE-FORM HOUSE

During the destructive days of World War I, architect Antonio Gaudi was busy developing a new, curving, free-form architecture in Spain. At the same time in Dornach, Switzerland, architect Rudolf Steiner was independently lending metaphysical credence to curving free forms in his famed Goetheanum, erected in that city. At the close of the war, a group of architecturally disillusioned German designers formed an intimate discussion group to explore the problems involved in establishing an improved house design. Communication with Gaudi and Steiner was sought by the group, and a round robin correspondence began which lasted for some twenty years. Mendelsohn, Kiesler, and Finsterlin proceeded from these ruminations to develop an entirely new, free-form building style, while most of their original group were wooed into the more popular International Style.

In more recent years, a group of young architects has been building from what was learned by these German pioneers. Foremost among these are the Italian Leonardo Ricci; the Americans Paolo Soleri and Bruce Goff; the Brazilian Oscar Niemeyer; the Mexican Juan O'Gorman; the Frenchman Andre Bloc; and the Britishers Hans Hollein and Walter Pichler.

106

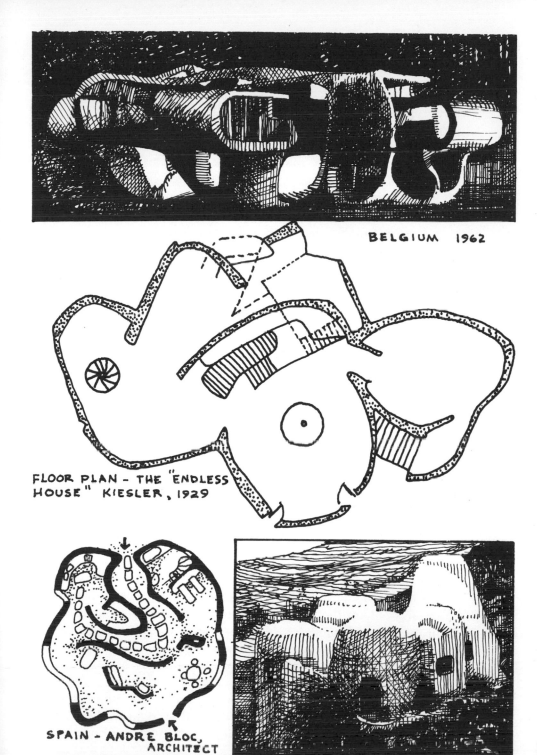

BELGIUM 1962

FLOOR PLAN - THE "ENDLESS HOUSE" KIESLER, 1929

SPAIN - ANDRE BLOC, ARCHITECT

11.1 FREE-FORM BUILDING

There is much about the free-form house that is applicable to owner-builder construction. This building style is not merely an art form or a radical return to nature, but it emanates from a life lived in dedication to fundamentals rather than one which is harnessed to mechanized equipment and interior decoration. The final form and shape of the house are reasonably determined by the usual height, width, and depth measurements that are required for the necessary functions of eating, sleeping, living, and working. Well-defined, functional areas can be closed off from or opened up to other areas, composing one continuous space. Finsterlin spoke of this "new house" as being organic, "the giving and receiving symbol of a giant fossil mother body." A person inside such a building would be as if inside an anatomy, wandering uninterruptedly from organ to organ.

Designers of free-form homes feel that there should be a greater independence from our ever increasingly automated way of living. Through architecture, these contemporary designers seek to encourage a more natural way of life among people. Kiesler speaks of orienting his house designs to simple, direct, and healthful principles of living, where work can also be recreational.

While the traditional house is post-and-beam construction, the free-form house differs from this standard with its continuous shell construction. The traditional, solid, opaque floor cuts through a free-form building like a pathological diaphragm. Designers of free-form buildings have, however, invested much thought to the elimination of flat, horizontal planes in the areas of peoples' movement. The floor of the free-form building curves at the rim into the wall and the wall curves, uninterruptedly, into the ceiling. Kiesler proposed a glassy, transparent floor where "the bare foot will caress the floor sculptures with every step, bringing new life to the neglected tactile sense and enriching the reactions which nowadays reach the level of our consciousness only as the grossest fragments, instead of as inconceivably delicate and pure melodies of the material world by which we are surrounded." Each sleeping section of a free-form dwelling is designed as an individual living room, with private bathing area included in each individual living quarter.

There is a continuity of structure in the curving free-form building, which makes it less erratic than stiff and straight-walled, angular constructions. Conventional structure is articulated from separate, stacked entities as compared with the simple, continuous expression made by surfaces of the free-form composition. Angles break the flow of the eye, but curves lead smoothly from one surface to another. There is also significant structural advantage in a continuity of structure. Forces applied at any point in such a structure will be distributed in an equal flow throughout the building. Obviously, these forces are less apt to flow at right angles to one another—which appears to be our expectation of standard frame or masonry construction.

Only the more serious and seasoned owner-builder should attempt to build a curving, free-form structure. To build in this manner, one must develop a knack for spreading concrete on curving forms, concrete plaster being the material best suited for this work. Regular poured concrete is too limited for use in this work because of the straight-edged formwork which is required with the use of this material. Plaster cement is a plastic material which virtually eliminates elaborate formwork. Plaster is applied to a reinforcement material which is, itself, used as a form, like a sculptor's armature or framework.

A lime coating was used to plaster coat buildings in earlier civilizations. Today, hydrated lime is used as a plasticizing agent to increase workability of this cement-based mortar, although diatomaceous earth and fireclay (mortar creme) are better to use as they are less harmful to workmen's skin. Gypsum and Portland cement are the two binders most commonly used in plaster work. Plastic cement is preferable to common cement for the mortar base since it is water repellent and spreads more easily. To achieve quality finish work only clean, well-graded plaster sand should be used. Coarser aggregate, when available, should be used in the base coats. Vermiculite and perlite aggregates are used when a lightweight acoustical effect is desired.

When plastering is done against a rigid surface, a metal stucco netting with large openings should be used so that the plaster will be pushed through to the backing, completely embedding the metal mesh with this applied surface material. A special furring nail is used to attach the mesh reinforcement to the backing. It holds the reinforcement out about one-quarter of an inch beyond the backing, permitting plaster to be forced behind it. Where there is no backing in place, expanded metal lath or gypsum lath are used to build a surface receptive to plaster. Always, plaster should completely cover the metal mesh.

TABLE 11.1

PLASTER PROPORTIONING, MIXING AND APPLICATION

Type	Surface	First Coat	Hard Finish	Sand Finish
Gypsum (fire-resistant or light-weight)	Interior only	2½ sand, 1 gypsum. 2½ cu. ft. vermiculite 100 lb. gypsum weight	1 lime putty, ⅓ gypsum volume	1½ sand, 1 gypsum volume
Gypsum (high strength)	Interior or exterior	2 sand, 1 H.S. gypsum weight	1 H.S. gypsum, ¼ lime putty volume	
Keene's Cement	On first coat of gypsum. Interior only		¼ lime putty, ¹⁄₁₀ fine white sand, 1 Keene's cement weight	4½ sand, 2 lime putty, 1½ Keene's cement volume
Lime (quick lime)	Interior or exterior	1 lime putty, 1 cement, 3 sand weight	1 lime putty, 2 sand, ⅓ gypsum volume	1 lime putty, 2 sand, ¼ gypsum volume
Portland Cement	Interior or exterior	3 sand, ¼ lime putty, 1 cement	Same as first coat	Same as first coat

The correct proportioning, mixing, and application of plaster is essential for quality results from the use of plaster. The table included here will assist the owner-builder in his selection and use of the common types of plaster. With some trial-and-error experience, one can soon develop a faculty for spreading plaster. It is also important to become familiar with and to purchase the correct tools for plastering. These tools include a high-quality steel trowel for application; a lightweight hawk for carrying mortar; a wood float for gliding over the surface to fill voids and to level bumps; a darby for preliminary smoothing and leveling. These tools will be illustrated in a later chapter.

There are methods of free-form construction other than plastering over an armature framework. Architect Paolo Soleri demonstrated one method when he built his 25- by 35-foot desert home. A huge mound of earth was first piled and then scored with V-shaped indentations into which reinforcing bars were placed. Wire mesh was laid over the entire mold, after which a 3-inch-thick layer of concrete was poured. When the concrete was set, a small bulldozer was used to excavate the earth material under the resulting concrete shell, piling it exteriorly against the sides of the building. The concrete roof meets the desert floor on the two long sides, against which the excavated soil is packed, and the building's entrance is discovered 6 feet below ground level. Opposite sides of the building open into excavated patios, where cooler air sinks to mix with air cooling in shaded areas.

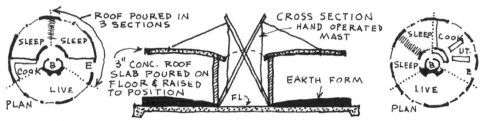

11.2 EARTH-FORM LIFT-SLAB

Concrete shells, walls, and roofs can be formed with what is called the lift-slab method. The major obstacle to building with the lift-slab technique is the necessity of devising a viable method for putting the usually very heavy components in place. One ingenious system involves casting sections in an earth form and then lifting the cured section into position with a hand-operated winch. Loose earth can be formed so that an arched roof (or curved wall sections) can be formed, longitudinally and transversely.

Soleri's desert house is not unlike Frank Lloyd Wright's berm construction method, designed in 1942 for cooperative homesteads. Wright pushed earth against outside walls, which provided the finished building with good insulation and protection from the elements. Use of this method of construction meant that outside walls below window-sill level did not have to be finished.

Earth forms have also been used in certain other types of lift-slab construction, a technique well adapted to free-form building. In 1956, Lawrence Carter invented a simple method for building conical, dome-shaped adobe houses. His experiments in Mexico demonstrated to him that anyone with even the most rudimentary skill in masonry construction could erect an all-adobe house. Total cost in those days

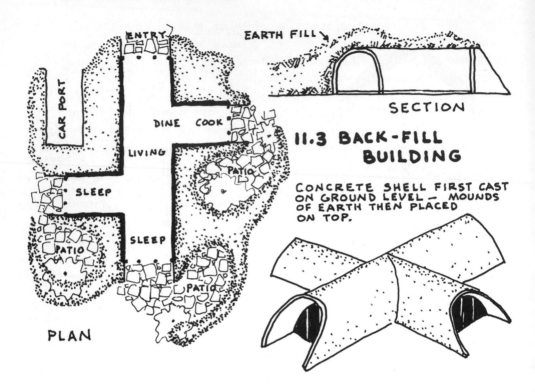

EARTH FILL

SECTION

11.3 BACK-FILL BUILDING

CONCRETE SHELL FIRST CAST ON GROUND LEVEL — MOUNDS OF EARTH THEN PLACED ON TOP.

ENTRY

CAR PORT

DINE COOK

LIVING

PATIO

SLEEP

PATIO

SLEEP

PATIO

PLAN

was 60 working days and $28.64 for materials. Carter's positioning machine was designed to serve as a guide for laying adobe walls. It consisted of a short, rigid arm revolving horizontally around a center pole. From the outer end of this arm ran two, longer, parallel arms, which extended back past the center pole to the wall on the opposite side. These arms could be raised and lowered, serving to support a shaped, wooden guide against which the adobe blocks were laid in domed form. The adobe walls were finally plastered inside and out.

Carter's earthen dome is the only dome known to this author that is adequately insulated, ventilated, and waterproofed. Three hundred square feet of living and sleeping area, however, may not be, by Anglo standards, much space for these functions, but a cloverleaf or a trifoil arrangement of several domes built around a large, enclosed commons or an outside patio could become an exciting owner-builder housing solution.

In Europe, a number of circular masonry houses have been built,

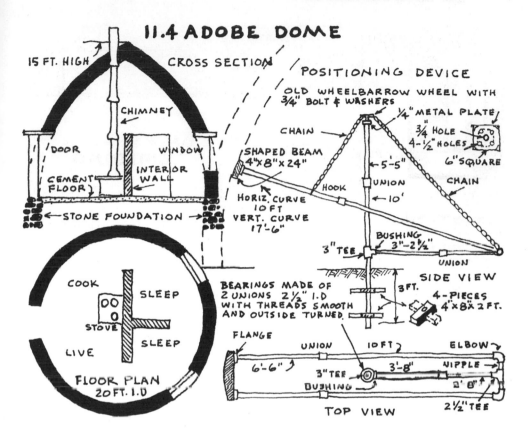

11.4 ADOBE DOME

CROSS SECTION

15 FT. HIGH

CHIMNEY

DOOR

WINDOW

INTERIOR WALL

CEMENT FLOOR

STONE FOUNDATION

HORIZ. CURVE 10 FT
VERT. CURVE 17'-6"

POSITIONING DEVICE

OLD WHEELBARROW WHEEL WITH 3/4" BOLT & WASHERS

CHAIN

1/4" METAL PLATE
3/4" HOLE
4-1/2" HOLES
6" SQUARE

SHAPED BEAM 4"x8"x24"

HOOK

UNION

5-5"

10'

CHAIN

3" TEE

BUSHING 3"-2 1/2"

UNION

SIDE VIEW

3 FT.

4-PIECES 4"x8"x 2 FT.

BEARINGS MADE OF 2 UNIONS 2 1/2" I.D WITH THREADS SMOOTH AND OUTSIDE TURNED

FLANGE

6'-6"

UNION

3" TEE

BUSHING

10 FT

3'-8"

NIPPLE

2' 0"

ELBOW

2 1/2" TEE

TOP VIEW

FLOOR PLAN

COOK

SLEEP

STOVE

SLEEP

LIVE

FLOOR PLAN 20 FT. I.D

but, except for Carter's modest offering, few builders in this hemisphere have been sufficiently interested in circular masonry wall construction to engineer a system that would make this design and construction feasible. One such system for building circular walls involves the use of a radius rod for determining building circumference. This rod is attached by a sleeve to a central, stabilizing pipe. The author experimented with this technique using a movable form and built rammed earth, concrete, and stone walls, all 6 to 24 inches thick. The principle is entirely sound. As long as the central pipe is set upright and plumb, the walls will, likewise, be plumb. The spacing radius rod, attached at its extremity to the form, runs to the central pipe. The circumference-spacing radius rod and the form are raised to the next higher, level position as the wall progresses in height. The

11.5 CURVED WALL BUILDING FORMS

lever mechanism of the form permits immediate release of the formed walls.

This same technique can be employed to build thin-shelled, circular, concrete walls. After establishing the vertical radius pipe, a concrete floor slab is poured, using a screed board that revolves around the radius pipe. To build the walls, a 5-inch-high by 3-foot-long aluminum form is filled with a concrete mixture that is dry enough to firm up, yet wet enough to allow the form to be moved forward to its next position as soon as it is packed. A continuous length of barbed wire reinforcement, running horizontally from foundation to roof, is used in each wall layer. Vertical reinforcement is achieved in the nature of the curved form itself. The form and the construction procedure are similar to those employed in the use of the Geiger horizontal sliding form, which will be illustrated in chapter twenty-two.

Instead of using a rigid rod to guide the slip form and to scribe the radius of a true circle, a flexible cable may be substituted and permitted to wrap around a centrally located, stationary plywood disc. An interesting spiral plan results. A prototype spiral-house building system has been developed by the author and is illustrated in Figure 11.6. The space between where the wall form begins and where it ends provides a practical and aesthetically interesting covered entry. It also provides an essential break in otherwise round-house monotony.

Personal experience building thin, curved, concrete wall sections,

11.6 SPIRAL WALL SLIP FORM

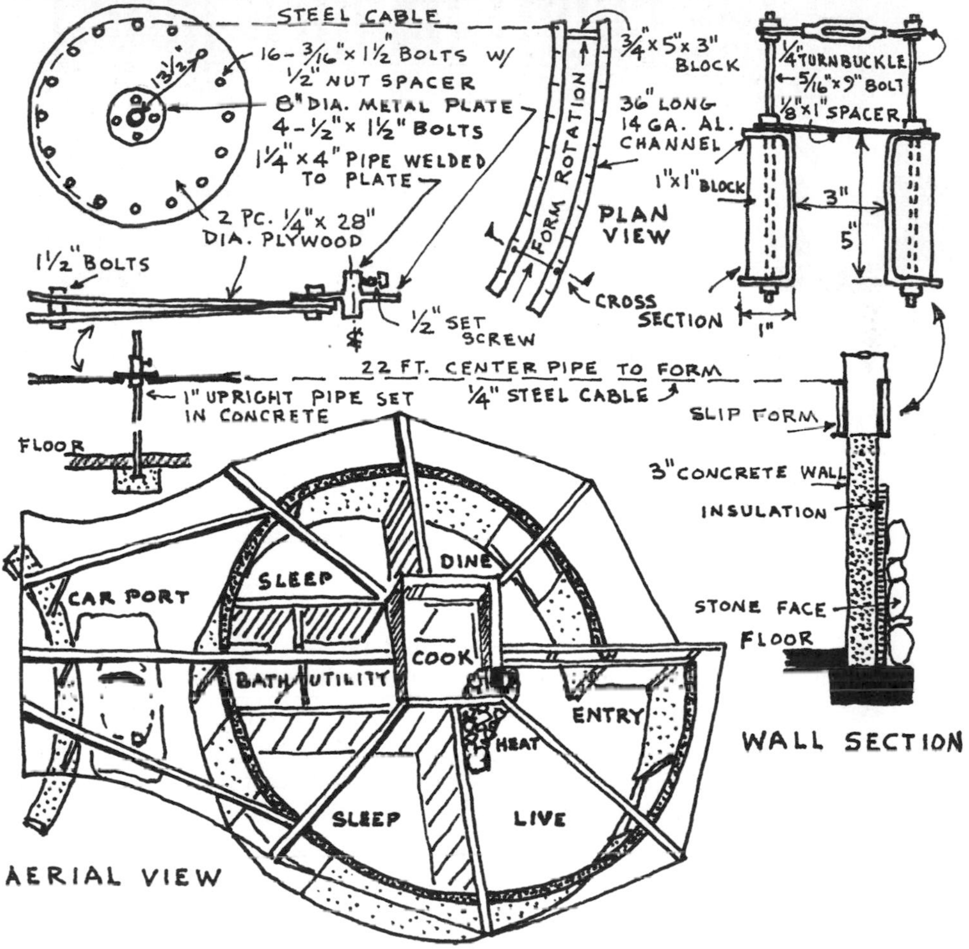

STEEL CABLE

16 - 3/16" x 1 1/2 BOLTS w/
1/2" NUT SPACER
8" DIA. METAL PLATE
4 - 1/2" x 1 1/2" BOLTS
1 1/4" x 4" PIPE WELDED
TO PLATE

2 PC. 1/4" x 28"
DIA. PLYWOOD

13 1/2"

FORM ROTATION

3/4" x 5" x 3"
BLOCK

36" LONG
14 GA. AL.
CHANNEL

1" x 1" BLOCK

PLAN VIEW

CROSS SECTION

1/4 TURNBUCKLE
5/16" x 9" BOLT
1/8" x 1" SPACER

3"
5"
1"

1 1/2" BOLTS

1/2" SET SCREW

22 FT. CENTER PIPE TO FORM
1/4" STEEL CABLE

1" UPRIGHT PIPE SET IN CONCRETE

FLOOR

SLIP FORM

3" CONCRETE WALL

INSULATION

STONE FACE
FLOOR

WALL SECTION

AERIAL VIEW

CAR PORT

SLEEP

DINE

COOK

BATH UTILITY

HEAT

ENTRY

SLEEP

LIVE

as little as 2 inches thick, excites this builder with the prospect of using this method for erecting shelter with a low first cost. Immediate occupancy is one of its benefits. Later, as time becomes available, interior partitions (if any) can be installed, and exterior walls can be insulated and veneered. Walls, being curved, are structural. They require few materials as they are thin, they can be built at little cost using local aggregate and sand, and they can be erected by inexperienced, owner-builder labor. All may be accomplished by using the unique and simple slip-form technique described above.

12 THE COURT-GARDEN HOUSE

> We often think that when we have completed our study of *one* we know all about *two*, because "two" is "one and one." We forget that we have still to make a study of "*and*."
>
> —Eddington, *The Nature of the Physical World*

As one advances from house planning to actual building, one soon discovers that the design of a house cannot be divorced from its structural purpose. The eighteenth-century sculptor Horatio Greenough recommended that man-made designs, like those found in nature, should express the function of the structure. This concept that "form follows function" asserts that the beauty of a building is relative to the degree to which it meets the demands of its function. A home with a high degree of living efficiency is almost always aesthetically pleasing.

One should not, however, confuse simple, efficient house beauty with the sort of rational purism that currently passes for modern architecture. In many quarters we find a revolt against designing dwellings with the rational materialism of the day in favor of designing buildings with more poetry and imagination. Architectural schools too often foster among students an academism of ascetic impoverishment which reduces the rising young architect, in Eric Gill's expression, to a "subhuman condition of intellectual irresponsibility."

The renunciation of the box house with its meaningless, decorative, applied art was asserted in 1915 by Spanish architect Antonio Gaudi.

116

Gaudi independently developed a flowing, sculptural, plastic quality in his building design, which remains unique to this day. He showed that superfluous design elements can be used if quality and measure control their integration. There is apparently a need in man for things that are not strictly necessary. In prehistoric times, man painted cave ceilings long before he knew how to build roofs.

Prior to cave dwelling, man lived in open spaces under the sun and stars. The cave was only occasioned as a protective shelter rather than as a habitation in which to abide. Man's craving for living out of doors continues strong within him. Open space, now considered essential to man's basic shelter needs, has long figured in the inner garden courts integrated into ancient Egyptian and Chinese constructions. The inner court of the Greek peristyle house evolved into the Roman atrium and eventually into the Spanish patio.

The first contemporary court-garden house was built in the 1930s in Germany by architect Mies van der Rohe. Despite its many advantages, very little was done with this architectural form until after World War II. A court-garden house offers areas of separation for

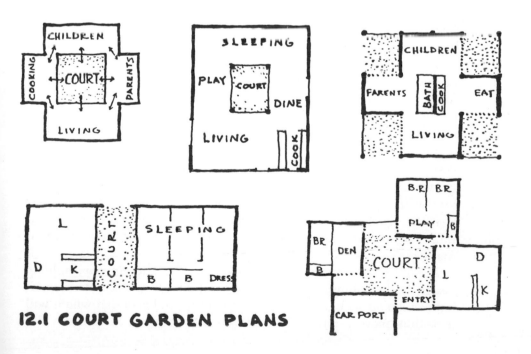

12.1 COURT GARDEN PLANS

various living functions, and it provides maximum privacy within the integral whole of the structure. Solar exposure and cross-ventilation are improved, and fewer outside wall openings and shorter spans offer major structural savings. Being inwardly directed, the court-garden house may have few openings or none on its exterior sides.

The court-garden house is merely one system for planning a livable and economical environment. Another system, the free-form, has previously been discussed. Both systems represent universally applicable design concepts founded upon basic construction techniques. One such concept is the design of the optimum circulation path, a universal constant in design and structural economy. A circulation path may be an enclosed hall or an unenclosed imaginary traffic lane. The Small Homes Council of the University of Illinois found that when the circulation area of a house exceeds 15 percent of the total floor area, the whole layout suffers. Beside being as direct and as short as possible, a main circulation path should be centrally located and should radiate to the various parts of the house. Ideally, one should be able to move to any part of the house from the main entry, without passing through any other sector.

A second universal constant in design and structural economy is the space-time relationship. Many of us continue to think in terms of life in a Newtonian world of three-dimensional, static space. A building exemplifies this static nature insofar as we have to walk around it at ground level to view its front, side, and rear elevations before we can fully comprehend its design and its structural components. A building, however, which offers continuously changing relationships between these components depicts a certain space-time quality, a dynamic relativity—motion in space. Compare Le Corbusier's roof-garden designs with the usual steep-pitched, shingled roof. The flat roof garden offers the viewer a new dimension, a new space relationship, a view from above as well as from below. The court garden offers a similar illusion of space in motion. There is a certain floating continuity in window transparency and its penetration of indoor-outdoor spaces.

The purpose for seeking this space-time quality in our building design is the intensification of our emotional responses. An owner-builder should seek to develop new relations in his design which will make him more emotionally sensitive and receptive. Much study has

12.2 ROOF GARDENS BY LE CORBUSIER 1930

been given to this subject by psychologists. At an American Institute of Architects' convention, Dr. Humphrey Osmond remarked that the most carefully designed buildings today are zoos. An animal will die if it is not properly fed and sheltered, but a human somehow learns to adjust. The emotional cost of this adjustment can, however, hardly be assessed, although it must be considerable. An ideal but nevertheless unchanging condition creates a dulling effect, which, when experienced in a comparable animal environment, will kill the heartiest beast. Freud has commented that irritants generate life. We really need the tensions and changes offered by contrasting and opposing spatial relationships.

Ectospace, mesospace, and endospace relationships offer contrasts in indoor-outdoor environment. Dr. Osmond distinguishes between sociopetal space, that which brings people together, and sociofugal space, that which keeps them apart. Contrasts are also formed between static space and fluid space. Fluid space exhibits motion which is eventually terminated in static space.

Given this analysis of space, the average owner-builder is no doubt more concerned with the tools and the methods for evoking in his design and structure these expansive emotions of discovery. It is important to know how to design, leading the observer through a sequence of experiences which reveal the dwelling as an organic unity—a pattern of rhythms, effects, and ordinal sequences, finally leading to a climactic repose.

The creative challenge in building design is essentially met by the development of scale and rhythm. Scale is dimension, which is relative to man's visual comprehension of an object in relation to his physical size. When we perceive our environment, we naturally use human-body size as a yardstick. The scale of median human-figure height has, therefore, been conveniently expressed as the 3-foot module. A module is a little measure, a division unit. Perhaps the one most important tool of design and structural economy is modular coordination. Modular layout describes the scaling or the sizing of a building in relation to its basic division unit. The 4-inch module is most commonly used, along with 16-inch and 24-inch modules. Four-foot modules are used as guides in gridwork design. All outside-wall lines and interior partitions are designed to fall on the grid module.

The correct ordering and proportioning of modules is essential to good design. For the creation of an intimate environment, one would, for instance, employ a smaller module than those used in the rest of the house, since reduced size gives an impression of coziness and a general sense of ease. Each individual element of the structure must somehow be related to the composition as a whole. Proportion is a composite of both function and scale and of materials and composition. Scale in a design is achieved through use of focus and contrast. The effective use of color can create a focus in one's building, while contrasts in light, shadow, and spatial sequence can contribute to the scale of one's design.

Rhythm design in building is virtually an unknown practice, even though the human capacity to respond to rhythm is innate and is the foundation of normal life. We create rhythm in our designs through juxtaposition of colors and textures, through the patterning of volume, through the succession of areas of varying light levels, through the spacing of free-standing columns. With these design aids, we can create a deliberate and measured rhythm, a beat exemplifying conscious dignity and poised assurance in our composition. Or we can introduce off-beat, secondary elements, such as openings or textured panels, to create a gayer quality in our design. Texture is used in contemporary design as an equivalent of ornamental detailing.

Geoffrey Scott has likened good architecture to frozen music. He also calls space liberty of movement. In our dwelling design we seek order, relationship, and integral structure, just as it occurs in inspired musical composition. An order of musical notes creates melody. The relationship of notes played produces harmony. The structure of musical elements played in a planned sequence produces a symphony.

12.3 COURT GARDEN

13 GROUP-LIVING SPACE

When planning group-living space, we seek, first, to satisfy our physical space needs with thoughtful placement of room furniture. We classify these furniture groupings as (1) primary conversation, (2) secondary conversation, (3) reading-writing-study, (4) music, (5) games. The resulting furniture arrangements define an interior view in manner similar to the way that window and door arrangements define an exterior view. A volumetric furniture sequence suggests itself with placement of the bulkiest furniture against solid walls, with furniture of medium-volume placed more casually about the room and with lighter furniture placed closer to window and door openings. An open-plan house design suggests use of a light, adaptable furniture. Low individual pieces may be sparsely distributed through the room.

Open planning offers an owner-builder a new release to a more satisfying living environment. Contrasted with free-flowing open planning, consider the studied approach to activity planning in the traditional, prescribed-sized, Victorian, box-shaped living room, designed exclusively to receive, entertain, and impress guests. The monotonous symmetry of box-shaped cells should be replaced with the spaciousness and flexibility to be found in the open plan. This open living arrangement adjusts to changes in demands from various occupants with varying activities by adjusting the interior space needed at different times of the day and of the year. There is provision for present activities, both passive and active, which are simultaneously engaged. This space also adjusts to different stages of family development. A virtually unrestricted inside view prevails in the examples of open planning illustrated below. Cooking and dining areas are only a few steps apart, and neither area is walled off from the rest of the group-living space. Quiet, passive functions are placed away from traffic areas in alcoves. By nature, we prefer the retiring feeling of an enclosure when, for instance, we choose to read.

Open-planning concepts serve as important tools to the modern architect, providing a means, not an end, to improved design practice. As is often the case, however, when the disciplines of one tradition are replaced, mere substitutes, not real alternatives, take their place. New directions too often necessitate a painful thought process and a major change in life styles. It may have been this notion that

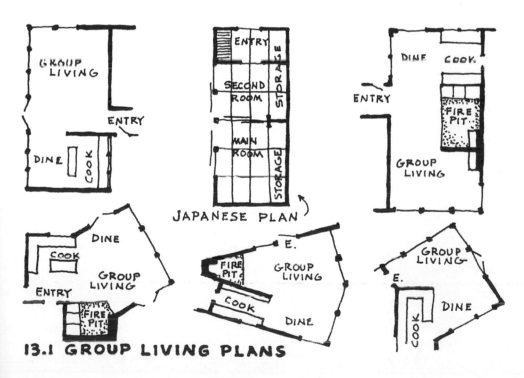

JAPANESE PLAN

13.1 GROUP LIVING PLANS

prompted Frank Lloyd Wright to remark, "In our country, the chief obstacle to any real solution of the moderate-cost house problem is the fact that our people do not really know how to live."

Open planning enables us to satisfy our physical group needs, but this is only one purpose for making spaces. We also design space to satisfy our impulses and to add richness to our lives. These emotional requisites are, too often, overlooked by architects and are seldom considered by owner-builders.

A surprise view or a changing vista in a room can add immeasurably to its charm. We should present the narrow-wide, the light-dark, and the high-low characteristics of a room. There should continually be a feeling of change in the places we inhabit. Wright spoke of filtering from one place to another. A hall, in effect, "filters" us into a room. The architectural treatment which we give to the hall, therefore, influences the emotional impact created through our transition from one space to another.

Entry-passage planning warrants our special attention and care. It is here that guests are first received, and it is here that we like to feel the first impact of homey welcome. A low, wide entry lends itself to a more gracious shelter suggestion than a tall, narrow one. A small, low-ceiling hall passage makes entry to an adjoining room a relief through a sense of movement and of spacious rest when one has actually entered the room. Hallway access and room openings should be grouped in proximity to one another to reduce traffic circulation directly across activity areas of a room and to discourage the division of the space into choppy sections. A good hall location is at the corner of a room rather than at the center of a wall.

Many alternative-housing concepts can be used to the economic and aesthetic advantage of the owner-builder. These ideas cut through established criteria and standards, and, for this reason, they meet with conservative resistance. Primarily, these freer, alternative design concepts tend to influence one's way of life, and this is sacred, carefully preserved territory for most people.

Alternative designing predicates that interiors will be planned for a complete range of activities and not just for singular, specific functions. There should be several possible locations for any single activity, and, conversely, one particular location should serve a number of different functions. We may choose to do hobby work, to read a book, to eat a meal, or to nap all in one room over a span of a few hours. Of course, noisy, messy activities should not conflict with quiet, passive ones. The best solution to this concern is provision for alcoves and nooks, which can be fully or partially closed while, at the same time, they are resolutely "linked" to adjacent spaces. A hallway doubles nicely as office space, sewing-laundry area, storage facility, or hobby center. Japanese house planning creates interior changes when movable pieces of furniture are alternately placed in storage areas and brought into rooms for use.

While alternative-housing planning offers one the occasional prospect of closing off some activities located in halls, alcoves, or nooks,

13.2 CONVERSATION PIT—POTTS HOUSE OAKHURST CA.
DESIGN BY AUTHOR—PHOTO, BOB BROOKS

the total is usually open into one, generous, continuous space. This one architectural feature is unvaried throughout the structure, and it includes all living functions.

Even the group-living fireplace area should be planned with alternative concepts in mind. The fireplace is traditionally erected in the center of an outside wall, due to fire hazard in early times. The mantel, raised hearth, and firebox opening are all carefully prescribed by former practice and executed with limiting symmetry. The couch is usually placed in front of the fireplace at a required distance of some 14 feet, which limits its use, intimacy, and privacy when the fire burns low. Seats can be placed at right angles to the fireplace and be backed against tall cabinets or bookcases for an inglenook effect. These seats are best placed at the left of the fireplace as one faces it for the same reason that right-handed people prefer doors opening to their right.

A more elaborate fireplace alcove can be created in the form of a conversation pit. An impromptu yet intimate atmosphere results from the dropping of this 12-inch recessed space. From a practical standpoint, in a relatively small area a number of people can be seated around the central focal point of the fire.

14 INDIVIDUAL-LIVING SPACE

Under the heading of "Individual-Living Space" we include all personal and private activities of a sleeping, bathing, dressing, and recreational nature. Our concern, as with family group-living spatial arrangements, is not so much for overall room planning as it is with the activities which will be pursued in various areas of the entire enclosure. General planning procedures should follow a sequence. First, a list in the order of importance should be drawn of one's personal, family, and social activities. Then, the conditions necessary for the pursuit and achievement of these activities may be deduced from an assessment of the available space, conducive atmosphere, potential efficiency, anticipated comfort, necessary furnishings, and requisite equipment. Next, one must group separately those activities that can be pursued together and those that cannot. They should be categorized in terms of time, place, sequence, and frequency of use. The determination of living-space requirements is based, in short, on the three-fold relationship of space (place), equipment (facilities and furniture), and atmosphere (physiological control of temperature, noise, and so forth). The successful determination of these three conditions will inevitably influence a healthful environment.

Health is man's ultimate need, and, as such, it should be the criterion for housing design. Yet, a dearth of research exists on this all-important subject. The most notable instance of work in this field was done in the Pioneer Health Centre at Peckham, London, from 1926 to 1951. Doctors at Peckham practiced preventive medicine and treated the whole person. They researched disease in its relation to one's living environment. Health is possible, they concluded, only when movement and flexibility are not impeded. In their requisites for building design, they considered as vital to a healthy environment those components that promote freedom of circulation, unimpaired visibility, and the flow of one space into another. Hallways were eliminated in the doctors' design proposals with the presumption that the whole building should function as circulation space. The primary purpose of the Peckham Experiment was to study function in healthy man. To realize his full potential, man, the doctors decided, must live in a fully free environment. Alternative design concepts are progressing toward considerable achievement of this goal.

A more detailed analysis of the functions of family living was conducted in 1941 by the Pierce Foundation. Interesting space and motion studies were made by this group and actual field studies of families were made in their homes. Family habits, attitudes, and possessions were evaluated, and physiological and psychological housing requirements were scanned. It appears that the home environment, in which we spend more than one-third of our life, may be one of the areas most neglected by designers and manufacturers. An individual's personal living space should offer something more than a boxy, 10- by 12-foot sleeping room, piled with manufactured goods.

One's personal living space should, first of all, be private. Everyone in the family, at one time during the day or night, should be able to get away from it all by withdrawing to his personal retreat. It is, therefore, advisable to locate one's personal living quarters some distance from the areas of group-living activity. Besides the reassuring awareness of the complete separation of these two areas with their opposing activities, we may also enjoy the prospect of eliciting a contrasting architectural experience from one's own personally inspired and executed microenvironment. A walk up a flight of stairs or through a long room to one's individual living area holds a certain attraction to some space-sensitive persons.

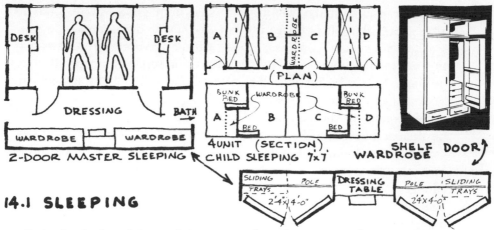

DESK

DRESSING   BATH

WARDROBE   WARDROBE

2-DOOR MASTER SLEEPING

(PLAN)

A   B   C   D

BUNK BED   WARDROBE   BUNK BED

A   B   C   D

BED   BED

4UNIT (SECTION)
CHILD SLEEPING 7x7

SHELF DOOR
WARDROBE

SLIDING TRAYS   POLE   DRESSING TABLE   POLE   SLIDING TRAYS
2'4"x4'-0"   2'4"x4'-0"

14.1 SLEEPING

It is the bed and its activity space that determines the size of the contemporary bedroom. An area of at least 10 by 10 feet is required to house a double bed, therefore a conventional three-bedroom house has about 400 square feet necessarily allocated to sleeping space. This area can be reduced to about one-half this amount by enclosing the bed in an alcove or a compartment, secondary to the room proper. This is not a new idea. Europeans for centuries slept in large, bed-containing cabinets. A modern equivalent of this bed-containing compartment might include the additional accouterments of controlled lighting, heating, ventilation, and sound-proofing. It has been suggested that the sides and ceiling of this compartment be lined with heat-reflective panels which would reradiate body heat during sleeping hours, obviating the need for confining blankets. Circulation of warm air under the floor of this chamber would supply sufficient heat, if the sleeping mattress were placed near or on floor level.

There are numerous advantages in sleeping close to the floor. Asians have, over the centuries, used floor sleeping arrangements. Comforters, folded and stored in a closet during the day, replace the bulky, massive bedding used by Western man.

Drawing 2.1 depicts the use of sleeping lofts—possibly the most sensible and inexpensive sleeping solution one could include in his new home. Loft sleeping should include outdoor decks for summertime sleeping use, for, without some outside access, a loft can, indeed, become a stifling environment. In summertime a raised sleeping level, a loft or a deck, catches cooling breezes. In winter living-area warmth naturally rises to make evening loft occupancy comfortable. Like an attic space, the loft helps to insulate the lower living area.

There is a general consensus among alternative-housing builders that the bathroom should lose its identity as a separate room. It was, perhaps, Le Corbusier who first altered the strict division between bedroom and bathroom when he, in 1929, placed the bathroom in the same room with the bedroom. He felt that the bathing area should be designed as a luxurious adornment to whatever room it occupies.

A bathing lounge, therefore, is presented herein as a desirable bathing arrangement. Complete with sauna, sun terrace, and cold plunge, these activities would be necessarily separate from a sink and toilet which would be secluded in one's own personal dressing area. The actual bathing lounge, however, could be planned in conjunction with sleeping areas and would function as a place for community bathing.

There is a current trend to place laundry activity in the bathing area. This makes more sense than placing laundry facilities in or near the cooking area. Most laundry comes from sleeping areas and is returned there after laundering for storage. Of all the rooms in the house, the bathroom is least likely to be upset by laundering activity during hours normal to that operation. Plumbing, hot water, and high-humidity room finishes are established elements of the bathroom-lounge.

Much work needs to be done to design more accessible and ample storage facilities for the individual living area. One should list all items to be stored. These items should be grouped according to their least and their most frequent use. Items which require special provision because of their weight or their size should be listed separately. Detailed closet and cabinet storage units can then be designed.

The usual closet has area that is impossible to use because sliding, folding, or swinging doors hamper inside visibility. A recent improvement in wardrobe design has furnished us with the shelf-door wardrobe arrangement, which provides us with more convenient storage space. Closet doors should open the complete facility to full view and immediate access. Customary drawer-faced cabinets conceal inside contents and often require the opening of many drawers to find an article. With shelf-door closet openings, sliding trays make article-and-clothes hunting a less tiresome and disagreeable proposition.

15 COOKING AND DINING

In chapter thirteen, "Group-Living Space," open planning and flexibility were discussed. These organizational concepts enhance and free
space for its functional aesthetic appreciation. Chapter fourteen,
"Individual-Living Space," emphasized the need for that spatial
arrangement which generates physical and psychological health. This
present section on the design and organization of cooking and dining
areas includes a consideration of aesthetic and healthful functioning
of those areas, and it, additionally, reviews some ideas which concern
human engineering, that is, engineering for human use. Cooking appliances, for instance, are designed for optimum efficiency, measured
by their comfort, safety, accuracy, and speed of performance. The
house we build to contain, among many things, this skillfully designed
machinery should be designed for the same high level of performance.
Many physiological work studies brought forth in England and in the
Scandinavian countries have been done to determine housing-
performance needs. Designers in these countries have engineered
equipment and housing to meet human requirements. Their considerations have taken into account: (1) the psychological aspects of
housing which have been conditioned by tradition and social pattern-

ing, (2) the physical impact of solar orientation, view, indoor climate, air circulation, and sound insulation, (3) the human engineering which involves a person's convenience arc, that is, one's height, reach, motion pattern, and space needs.

In contrast to this human engineering approach, a University of Illinois Small Homes Council survey of kitchens in over a hundred housing developments found that 90 percent of these cooking areas had inadequate base cabinet storage, 77 percent had too few wall cabinets, and 67 percent had restricted counter space. From the standpoint of human engineering, there are five requirements for an optimum food-preparation work center: (1) adequate activity space (2) adequate counter space, (3) adequate equipment space, (4) adequate storage space, (5) an arrangement of all these areas for maximum efficiency. Obviously, few home builders achieve the development of a truly efficient work center.

15.1 COOK-DINE ARRANGEMENTS

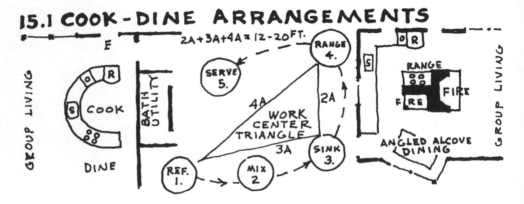

Motions take time, so that, when one is designing a cooking area, the first question to ask is, "Where is the best location for what?" To answer this we must analyze the work to be done. For a right-handed person the cooking sequence is from right to left: store, clean, mix, cook, serve. For each of these functions, we determine the needed equipment and supplies. Equipment should be arranged in the sequence required to do the job and at heights related to body use. Cooking research at Cornell University established that a triangular relationship between refrigerator, sink, and range, with any combination of distances between these fixtures of from 12 to 20 feet, would be highly efficient and would save energy. From the standpoint of human engineering, a kitchen sink should be 3 inches higher than the standard 36-inch counter height, and the mixing center should be 4 inches lower than the standard, which is 32 inches high. Physical strain results when the average American adult reaches into a storage cabinet that is lower than 20 inches from the floor or higher than 60 inches from the floor on which he stands. Strain occurs while ducking to avoid being hit by an open upper cabinet door when an abrupt, negative, backward angle of bend is made by the body. Sliding door cabinets are preferable. Sufficient floor space should be provided for working in front of and for passing between each element of the work center. Finally, planning for cooking facilities should take into consideration such necessary features as lighting, acoustics, heating, and ventilation.

Alternative-housing cooking design provides material storage at the point of first use in each major cooking activity center. The major cooking centers are designated as refrigerator, sink, mixing center, range, oven, and serving area. Wall and base cabinets for each of these activities should be the same length, about 4 feet for each unit, excepting those for the sink, which should be about 8 feet long. The space beneath an upper wall cabinet creates a convenient place for the installation of indirect lighting, which shines directly onto the work counter. The usual center-ceiling light fixture sheds illumination where it is least needed.

Much research on cabinet storage has been done in recent years. Perhaps the most noteworthy is the work done at Cornell University, whose research in this area has already been mentioned. A type of swing cabinet has been suggested by the University group for kitchen use. This is a compact cabinet made of sections that swing open like a book. Storage is one row deep, making each item easy to see and to grasp. Only the item wanted has to be removed. Racks on storage doors are a sensible device for storing small food items, as well as for storing small cooking utensils, spices, and so on. Swinging-door base cabinets with the usual stationary shelves should be avoided. Shallow, pull-out trays and drawers give better visibility and greater ease for reaching contents. Heavy pots and pans are brought into full view and easy reach by pulling a tray forward. Vertical drawers are particularly satisfactory below the sink, where the often-used dishpan, dish drainer, and brushes may be hung on hooks. A similar vertical drawer beside the range is handy for flat pans and pot covers. Vertical file partitions can be installed to advantage, too. Articles stored in these files are within easy reach and can be grasped readily. To avoid being hit by open swinging doors, overhead cabinets should have sliding doors whenever possible. They do not offer as full an exposure of contents as do swinging doors, but this possible disadvantage can be overcome by using glass-panel doors when necessary.

A poorly designed cooking area costs as much to build as a good one. The popular Pullman counter cuts across traffic areas and is too long for efficient use. The L-shape counter arrangement is better, especially when the range is located at the corner of the L, where

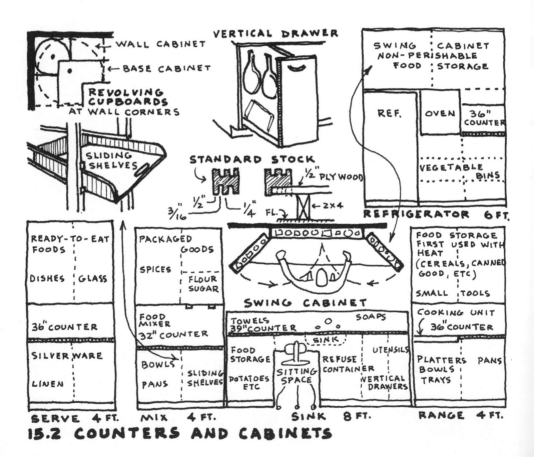

15.2 COUNTERS AND CABINETS

undisturbed cooking can be done. Probably the most practical and efficient cooking arrangement is the U-shape plan. A variation of the U-shape scheme is a circular cooking layout, which was found to require only 70 feet of walking for meal preparation (see Figure 15.1). The same meal prepared in an L-shape cooking area required 245 feet of walking.

Alternative-housing design chronicles fresh, new approaches to cooking-dining organization. The traditional window-over-sink orientation, for instance, is now considered obsolete for proper lighting of that work area. Work areas are best lighted by clearstory or by skylight. The dining space or mixing center might utilize some window exposure. In a good plan, the cooking area is convenient to the garage as well as handy to the front entrance. However, entrances to the cooking area should be grouped at a wall corner to minimize through traffic.

Another alternative-housing approach to opening up the cooking area is the combination of cooking-dining-family room functions. This single-space grouping does not isolate the cook from the social activities of others. The formal dining room of the 1920s has dwindled from room to dinette to nook. Actually, the dining room can function better as a secondary group-living area, with placement of the dining table in a windowed alcove or in a firepit of its own. The table should be as close as possible to the food-cooking and food-service area. It is also desirable to locate the table near the sink for simplified cleanup. When food service and cleanup are separated from the dining area, a utility cart is a useful device. The Cornell investigators designed a cart for this purpose that holds service for eight people.

16 INSIDE YOUR HOME

> Modern architects have been harping continually on what is different in our time to such an extent that they have lost touch with what is not different, with what is essentially the same.
>
> —Aldo van Eyck

Exciting changes are taking place in alternative-housing interior design. Lao-tse has been quoted as saying that the important part of a building is not its walls or roof but its empty spaces. For purposes of discussion, we must differentiate between inside space and outside form. Frank Lloyd Wright said that what happened on the outside of a building occurred because of what was happening on the inside. Houses should be designed essentially around what we do in them.

Let us recognize, first, the animal nature of man. We design to satisfy needs or, more explicitly, we design to achieve comfort. Heretofore, this book has been directed to aspects for securing physical comfort. Something, now, must be said for the equally vital concept of psychological comfort. The overall effect of interior space upon one's senses and one's consciousness defies complete definition, but it can be partially analyzed. Sensory reactions to room environment can be relaxing or invigorating, or they can be disturbing.

The owner-builder who expects to create a pleasing interior environment should not take himself too seriously. His tone should be one of relaxed informality. He should remain experimental, and, above all, the creative living experience should be fun.

Architect Robert Venturi claims that the best architecture is not symmetrical or balanced, nor is it clean and simple, logical and formalized. According to Venturi, to achieve a vital and timely reality, the architecture must contain what traditionalists call confusions and distortions. It must be complex, contradictory, and ambiguous and contain downright error in concept and execution.

Our reaction to an enclosed space is a reaction to its size, shape, lighting, color, openness, and so on. To a space-sensitive person, a long and seemingly endless corridor is disturbing. Anxiety is created for some when this space creates distortions of perception. A space that does not have a clearly defined size or shape can produce a feeling of insecurity. For such persons, a space should be immediately comprehensible.

The new look in building interiors is one of *boldness* in lighting and in color. Lighting is no longer thought of as merely illumination. Rather, its contribution to psychological relief and to room atmosphere are its prime value. Spot lighting is employed to highlight or focus attention on an area. Recessed down-lights create a sophisticated wash of modified lighting. Table and floor lamps furnish portable, selective lighting.

On one hand, we seek to produce a psychologically stimulating environment, while, on the other hand, we choose to subdue certain dominant elements of our home life. The environment we choose to create should complement, not compete with, our social contact. We all know the irritation of competing with so-called conversation pieces.

The competing needs for genuine social contact and for privacy in our lives deserve the utmost consideration. In either case, a satisfactory experience is possible by raising or by reducing barriers in our environment. Alienation, aggravated by poor spatial organization, is relieved by wise spatial planning. Paths for circulation should be laid out to provide people with access to all activities. A select work space should have at least a visual relationship to the total space in which it is placed. Face-to-face personal contact can sometimes be aided by the thoughtful use of adjustable furniture.

Some furniture items can be advantageously mounted on wheels, but wherever possible one should use built-in furniture pieces. They go far toward eliminating furniture clutter. Such a consolidation of furnishings is an attractive idea to people who are unemcumbered by

SECTION

STORAGE

ENDO BED

COUCH RAIL

FIRE TABLE TRACK

MESO

ECTO

FLOOR PLAN

GROUP LIVING

PRIVACY – STUDY

COOKING – DINING

SLEEPING – DRESSING

16.1 POLYFUNCTIONAL ENDO-SPACE

conventional trappings. Dispensing with the usual, traditional furniture paraphernalia implies economic saving as well as personal release. There is major economic advantage in creating minimal rather than fulsome interiors. The elimination of interior nonessentials maximizes room dimensions.

The development of living centers consists of clustering equipment and furniture into various portions of a room. Systems furnishing synthesizes and consolidates furniture and equipment and creates a fresh, revolutionary view of the whole furniture concept. In a particular living center, the furniture may very well consist of movable trays or platforms. They can be wheeled, slid, or taken apart into pieces and laid out in a variety of ways. One polyfunctional living center may thus become at various times a living, dining, sleeping, or study area.

Alternative-housing furnishing contrasts with current furniture-grouping practices in about the same way that mobile-home furniture use contrasts with proposals for boat interiors. A boat is designed to utilize total space; amenities are built in. The space in a boat is small but highly integrated. The mobile home is small in space also, but it is, nonetheless, furnished with the usual assortment of standard-sized furniture pieces and appliances. The prefabrication of the shell of this moving shelter does not transfuse to the design of its interior.

Although open planning is idealized in this book, there is danger in overdoing it. One of the more gratifying senses is that of privacy.

This is visual and acoustical privacy as well as spatial, tactile privacy. Aldo van Eyck interjects some cautioning words to say:

> We must break away from the contemporary concept of spatial continuity, and the tendency to erase every articulation between spaces; i.e., outside and inside, between one space and another. Instead, the transition must be articulated by defining the in-between places which induce simultaneous awareness of what is significant on either side.

An endo-, meso-, ecto-space design exemplifies this concept. The inside of our homes should be designed to serve many functions and these functions should be expected to change. Flexibility is the key to the awareness and gratification of our senses, our moods, and life programs. We need to sit in different postures and positions at various times. During certain periods of life, we require diverse places and arrangements for eating and sleeping.

When systems furnishing is not used, a room's floor should be kept bare. Many of the room's things can be stored out of the way in convenient wall-storage cabinets. Rooms are too often centers for one's display of possessions. If passage areas are expanded into usable, space-saving storage facilities, the size of adjoining rooms can be reduced, saving construction costs. Costs are further reduced by elimi-

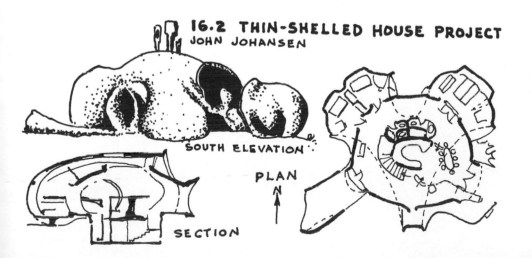

16.2 THIN-SHELLED HOUSE PROJECT
JOHN JOHANSEN

SOUTH ELEVATION

PLAN
N

SECTION

nation of reveals and molding trims. Flush, frameless window and door openings and broad expanses of plain surfaces also contribute to savings. A poorly designed interior cannot do permanent damage to a well-designed house, but it can surely ruin it for the duration of its occupancy.

Frank Lloyd Wright said that corners put an end to space. This idea is worthy of contemplation. It just may be that some of the spatial features mentioned above can be achieved in a straight-walled structure only with the greatest difficulty and compromise. This writer personally feels that a rectangular or a cube-shaped room is unimaginative, confining and depressing. Circular, curvilinear, or organic spaces, though they may seem novel or difficult to construct, feel right in their own pure state, in their simple, undesigned form.

17 ADOBE BLOCK

Of the next dozen chapters reporting on the materials of construction, it is not by accident that this first chapter particularizes elements in the production of sun-dried adobe block. Soil construction using adobe blocks was in use by the end of the Neolithic Age. Traditionally, it is one of man's most popular as well as his oldest building material. Earth-block technology was developed early in man's history, and to this day it has changed little. A natural asphalt stabilizer, for instance, was employed by ancient Babylonians to improve the weather-resisting qualities of soil walls, much as we do today.

It is the owner-builder who can best appreciate the value of earth-wall construction. No commercial or private-interest group can be found to extol its merits, for there is nothing in bare earth to sell. However, government-sponsored housing projects, such as were erected in this country during the Depression, have constructed hundreds of earth homes and public buildings, and, during one year in India, 4,000 permanent earthen homes were built for displaced persons.

Despite their real value, earth buildings are, as we know, the exception rather than the rule. In our own land of freedom, it is quite uncommon to find a building ordinance which will approve an individual builder's earth-wall construction. Who is there among us so naive as to not recognize the intention to influence building-code requirements by building materials manufacturers? One effective lobby by the American Lumbermen's Association to the U.S. Department of Commerce succeeded in a complete shelving of the earth-wall program.

With unquestioning loyalty and unbelievable ignorance, building inspectors say "No" to earth-wall construction in this country, yet, according to structural tests made at the School of Engineering, Christchurch, New Zealand, a foot length of soil-cement wall, 8 inches thick, will carry over 21 tons at failure. The weight of each lineal foot of wall, 8 feet high, is approximately one-fourth ton. That leaves $20\frac{3}{4}$ tons for roof weight and safety. The Australian Commonwealth Experimental Building Station found that the compression strength of an adobe block is in excess of 25 tons to the square foot. Our own Farm Security Administration claims 33 tons per square foot compression strength for these blocks. This is actually about ten times the strength needed for conventional roofing weight. A complete report on structural heat-transfer and the water-permeability properties of

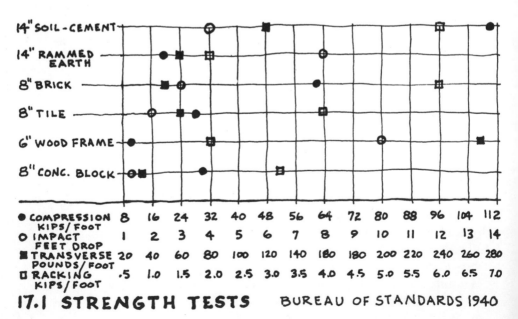

● COMPRESSION KIPS/FOOT	8	16	24	32	40	48	56	64	72	80	88	96	104	112
○ IMPACT FEET DROP	1	2	3	4	5	6	7	8	9	10	11	12	13	14
◼ TRANSVERSE POUNDS/FOOT	20	40	60	80	100	120	140	160	180	200	220	240	260	280
◻ RACKING KIPS/FOOT	.5	1.0	1.5	2.0	2.5	3.0	3.5	4.0	4.5	5.0	5.5	6.0	6.5	7.0

17.1 STRENGTH TESTS   BUREAU OF STANDARDS 1940

various earth-wall constructions was issued in 1940 by the Bureau of Standards, and this report is reproduced in Figure 17.1.

In areas of earthquake or high wind, special regard must be given to earth-wall engineering. It is best to keep the building compact in design and to build the walls on a continuously poured, reinforced concrete foundation. In earthquake country, it is recommended that one build walls thicker than in nonearthquake areas, perhaps 18 inches thick. Horizontally placed, reinforcing rods or wire should be embedded in the mortar between blocks to help tie walls together and to reduce shrinkage and cracking. Finally, in order to adequately tie the walls together and at the same time to reinforce the top of the earth wall against loads from the roof, a reinforced, concrete-bond beam is poured in place on top of the completed earth wall. Ideally, the bond beam is designed to serve as a lintel for support of door and window spans as well as to act as a perimeter tie. Ceilings and roofs, preferably built of lightweight materials, should be anchored to both the ends and the side walls and should be constructed to serve as diaphragms which resist lateral distortion.

None of these reputable construction practices are acceptable to California building-code-enforcement agencies. Accordingly, to build with earth one must use the earth-nogging method, in which earth material is merely filled between wood, concrete, or steel framing members. Regardless of restrictive building codes in this country, however, years of trial and error with earth constructions around the world have culminated in newer systems for machine-compacted blocks and stabilized soil-cement constructions.

Generally speaking, there are two classes of earth-wall construction: puddled and tamped. In the former instance, a molecular aggregation of earth particles is achieved within a liquid medium, the puddling process. In the latter instance, the earth particles are compacted by the use of compression—the tamping process. Adobe blocks are puddled (wet), while rammed earth is tamped (moist).

Whether puddled or tamped, the earth may be stabilized or unstabilized. The common stabilizer used in the puddling process is bitumul, while that employed in the tamping process is usually ordinary cement. Earth may also either be precast into blocks or cast into forms in situ. Walls of the owner-built home can, therefore, be puddled or tamped, stabilized or unstabilized, precast or made in situ. One's choice of system depends upon a host of factors, such as design, type of soil, facilities, equipment, and the availability of workers.

It has been found that practically any soil can somehow be used in earth-wall construction. A soil that proves unsuitable for building by one method may be entirely satisfactory for another. For instance, a soil used in a poured adobe wall may shrink and crack, but the same soil may prove satisfactory for adobe blocks, since the blocks are preshrunk before being placed in the wall. Also, in tamped-earth construction, consolidation of earth particles reduces shrinkage.

The sandy-clay-adobe soil of the arid Southwest is usually considered ideal for puddled-earth-wall construction. It is at least 30 percent sand and not less than 50 percent clay and silt. The clay provides compression strength, and the sand reduces shrinkage and cracking by lowering the absorption of moisture. Adobe blocks are molded from clay that is in a plastic state, often with a moisture content as high as 30 percent. A moisture content of from 15 to 18 percent is considered optimum. Straw binder is sometimes used to reduce cracking in unstabilized blocks. The straw, cut into lengths of from 4 to 8 inches, is evenly distributed throughout the mass. About 150 pounds of straw may be used to make 1,000 4-×-12-×-18-inch blocks. Very little fiber decomposition occurs in adobe block. Blocks over 100 years old have been found in the Southwest containing dried grasses in such perfect condition that the species could be identified.

Interestingly, it is from the science of road building that we have learned the most about the way soils behave. Road builders have found ways to stabilize those soils which would, otherwise, be unsuitable for building purposes. The original purpose of bitumul, or asphalt emulsion, was for sealing road beds so that one finds this material readily available from road-paving companies. After asphalt emulsion has been added to the soil, it separates back into pure asphalt and water, leaving the asphalt as a film on the grains of soil. An emulsion that slowly separates back into asphalt and water is called a slow-breaking or a slow-setting emulsion. It is best suited for earth walls, because it separates only after being thoroughly mixed into the soil. The American Bitumuls Company produces a stabilized hydropel, ideally suited to adobe-brick manufacture. Costs for fifty-gallon drums are about 30 cents a gallon. It would, actually, be less expensive to make an unstabilized block and to waterproof it on the outside. According to tests conducted by the U.S. Bureau of Standards, stabilizing blocks does not appreciably increase their strength. Furthermore, the insulation value of the block is reduced when the density of the block is increased by the stabilizing process.

The first important step to take when contemplating an earth-wall house is to test the soil that is available at or convenient to one's building site. Again, road builders can show one how to determine soil types and, accordingly, how to determine the type and the amount of earth stabilizer to use. The simplest test is merely to pick up a handful of dry earth and to rub it between the fingers. Sand particles are gritty to the touch, while silt and fine particles adhere closely to the skin and have a silky feel. These physical properties, which can be felt and seen, are largely responsible for the strength and durability of walls, or, conversely, they may be responsible, due to the movement of moisture through their mass, for the susceptibility of walls to crack. Warping and cracking are characteristics of unstable colloidal clays, which readily take up moisture. Sand is known to reduce this shrinkage, but excessive amounts of sand also prevent proper bonding. Clay bonds the coarser granular minerals together, but it has the unfortunate characteristic of being a thirsty material. Clay swells as it absorbs water. As it releases moisture it cracks. Then, too, a soil high in silt content produces a wall which readily erodes. A wide variety of soils can be used in earth-wall construction. Satisfactory earth can be found almost anywhere, preferably in the subsoil region, three feet or less below the top soil.

In another simple test for determining exact sand-clay proportions, fill a one-quart mason jar one-quarter full of the soil sample. The earth should first be screened through a No. 4 sieve, 6 squares per inch. The jar is then filled with water, and a spoonful of common table salt is added to speed up the settling of the clay. Some road engineers prefer to use a 10 percent solution of ammonium hydroxide to speed this settlement. The jar should then be agitated thoroughly and allowed to settle for one hour. The sand and clay will settle in successive layers, the bottom layer being sand and the top layer being clay. Measure the height of the sand and divide it by the total of soil settled in the bottom of the jar. This will give the percentage of sand in one's sample.

Proportions are variable and must be worked experimentally before certainty can be attained. It may, therefore, be correct, but it is certainly not sufficient, to say that any earth containing at least 50 percent clay and silt is suitable for adobe blocks. The amount of moisture required to bring this earth sample to a suitable consistency for molding may be 16 percent by weight, but then again this amount of water depends upon the clay-sand proportion in the sample. Neither can one assume that a 50 percent clay-silt content always requires 10 percent asphalt emulsion by weight, although that amount seems to be optimum for these proportions.

In general, soils with high sand content require less stabilizer. A soil composed of 70 percent sand needs 5 percent or $\frac{1}{3}$ gallon of emulsion per cubic foot. The only satisfactory way to determine soil, water, and stabilizer proportions is to make a number of sample blocks and to evaluate them discerningly. Each block should contain differing quantities of stabilizer, ranging from the lowest amount that seems to produce satisfactory results to the greatest amount that one can afford. Evaluation tests start with a determination of water content. If the blocks slump or settle when the form is removed, the water content is too high. After the blocks are sun dried for a month, check for cracking. Open cracks indicate too high a clay content. If the blocks can be crumbled easily, the soil is too sandy.

A spray test should be performed to determine how well the block will hold up in a hard, driving rain. If the spray nozzle of one's hose is held 7 inches from the block at a water pressure of 20 pounds per square inch for 2 hours, only a slight pitting should occur.

The most important block test to be made is one that determines the crushing strength of the soil. A simple lever mechanism can be fabricated to facilitate the testing procedure. A sample block should be roughly twice as long as it is wide, and it should be loaded in the direction of its longest dimension. As shown in Figure 17.2, weight is first applied close to the block, then moved slowly outward to the end of the lever until the block breaks. A minimum of 250 pounds per square inch is essential for dry compressive strength.

Tests on blocks which have been immersed in water for a number of days can be even more revealing. Wet strength should be at least one-half of the dry strength. In areas where there is a lot of rainfall, the wet strength should be nearly equal to the dry strength.

Dampness at ground level and just above it is responsible for the damage that usually occurs to an adobe wall. This is particularly true

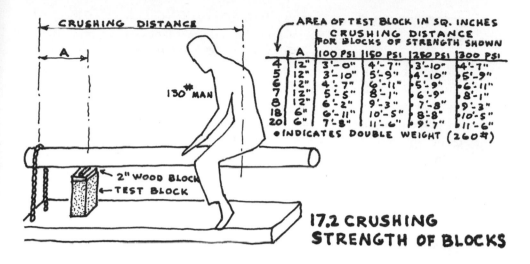

	A	100 PSI	150 PSI	250 PSI	300 PSI
4	12"	3'-0"	4'-7"	3'-10"	4'-7"
5	12"	3'-10"	5'-9"	4'-10"	5'-9"
6	12"	4'-7"	6'-11"	5'-9"	6'-11"
7	12"	5'-5"	8'-1"	6'-9"	8'-1"
8	12"	6'-2"	9'-3"	7'-8"	9'-3"
18●	6"	6'-11"	10'-5"	8'-8"	10'-5"
20●	6"	7'-8"	11'-6"	9'-7"	11'-6"

AREA OF TEST BLOCK IN SQ. INCHES — CRUSHING DISTANCE FOR BLOCKS OF STRENGTH SHOWN

● INDICATES DOUBLE WEIGHT (260#)

CRUSHING DISTANCE

A

130# MAN

2" WOOD BLOCK
TEST BLOCK

17.2 CRUSHING STRENGTH OF BLOCKS

in countries where it may freeze in the winter. Earth-wall protection must be provided either by stabilization, by application of a protective wall covering, or by the design of the structure itself. Stabilization is achieved through the introduction of an emulsified oil additive into the mud mixture. Protective wall covering is achieved by the application of plaster or other protective wall-surfacing material. Full protection of earth walls is ultimately achieved by design of wide roof overhangs, verandas, and so forth.

Whenever possible, a building design and wall system is recommended which does not require applied protective coverings. Even the best covering is subject to continual maintenance and occasional repair. Great care must also be observed in applying a protective covering, lest it fail to bond to the wall, allowing moisture to collect between coating and blocks, and it eventually disintegrates with the continual thrust of freezing and thawing.

The most common mistake made when applying protective finishes to earth walls is made when the finish is applied before the wall has completely dried out. Moisture in the wall, escaping through the finish, will soften the finish at the point of bonding, and it will fail in its purpose. Some sort of mechanical bond, such as wire mesh, is always recommended for the plaster coating of walls. Some of the more satisfactory methods for protecting the exterior of earth walls will be discussed in the following chapter.

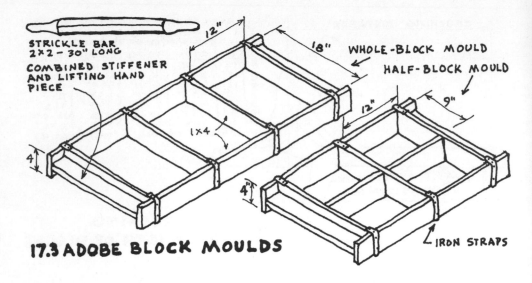

STRICKLE BAR
2×2 – 30" LONG
COMBINED STIFFENER
AND LIFTING HAND
PIECE

12"

18"   WHOLE-BLOCK MOULD
HALF-BLOCK MOULD

12"   9"

1×4

4"   4"

17.3 ADOBE BLOCK MOULDS

IRON STRAPS

Adobe blocks are made by placing wet mud in wood or metal boxes, called molds. A straight-edged piece of wood, called a strickle bar, is used to smooth off the top surface of the block, making it level with the mold. The Australian Commonwealth Experimental Building Station devised a set of wooden adobe-block molds that are as good as can be found anywhere. As soon as the adobe mud is kneaded into the corners of the mold, the mold is picked up and placed alongside the completed block for the forming of the next one. Blocks should be kept in their original position for at least 2 to 3 days so they will harden sufficiently for handling. Then, in a sheltered curing area, the blocks must be stacked on edge, with air spaces between, for about three weeks before they are used. Blocks with a high clay content should be cured more slowly.

Adobe-block production can be substantially speeded by using a large rectangular wooden frame one brick deep. The dimensions of its sides are in multiples of the breadth and length of the block. It is filled and packed with adobe earth and leveled off. Then the block sizes are cut into the mass of earth with a knife drawn along the straight edges of the frame.

Adobe mixing by hand is slow and tedious. Some kind of mechanical mixer can, therefore, be used to advantage. The best types are the pug, dough, or plaster mixers. Concrete mixers are generally not considered satisfactory unless modified with blades or baffles to break up the material as the drum revolves.

Blocks are held together with a mortar, which is made from the same mud used for making the blocks, giving both mixes the same coefficient of expansion. A small amount of cement can be added to the mortar, however, to make it set faster and to add to its strength. The mortar proportions found most suitable by the author are: 1 part cement and 4 parts sandy soil (by volume), in which emulsified asphalt is incorporated in the ratio of $1\frac{1}{2}$ gallons to each sack of cement.

The advantage of self-made adobe blocks is that one man can carry out the whole molding and building operation unaided. The owner-builder can make bricks in his spare time and store them until they are to be used. Soils of higher clay content can be used for making the brick, since any shrinkage will occur during drying, before the blocks are built into the walls. The manner in which earth-block walls are laid up will be covered in the following chapter.

18 PRESSED BLOCK

There is one fact about earth-wall construction that should be faced at the outset of any job by an owner-builder. Excessive amounts of earth material have to be handled. Two thousand adobe blocks are required to wall up the outside of a small house. This represents 30 tons of earth, 80 cubic yards, and several months of grueling labor making the blocks and then laying them in place. Few North American owner-builders can afford this investment in labor and time, while in house-needy, third-world countries there are few people who can even afford the capital outlay for this kind of dirt-cheap housing.

Industry has, therefore, devised tools with remarkably higher efficiency than former hand-pressed equipment for producing pressed block. First to be produced was the British Winget machine. It is a hydraulically operated block press powered by a small gasoline engine. It consists of a hand-rotating table having three operating portions, which include filling the mold, pressing the material, and ejecting the block.

A South African company redesigned the Winget and added a number of basic improvements. Called the Landcrete, this press can produce as many as 500 blocks per hour. A hand-operated version is also manufactured which will make 150 blocks per hour. One exceptional feature of the Landcrete is the interlocking block that can be produced, a tremendous aid for the unskilled block layer. A number of different sizes of blocks can also be made, including a hollow-corner block. With this special block it becomes possible to construct steel-reinforced concrete corners and piers.

Another South African manufacturer produces an interlocking block machine, the Ellson Block Master, which operates on a hand-operated lever system. The high lever ratio (500 to 1) produces a dense block at about the same production rate as the Landcrete machine.

In 1952, the Ellson Block Master was redesigned by the Inter-American Housing and Planning Center (C.I.N.V.A.) at Bogota, Colombia. Paul Ramirez, a Chilean engineer and the inventor of the Cinva Ram, worked for several years developing a device which would, in his own words, ". . . give families of small means a manual tool that will help them build the walls and floors of their houses." The specifications for this tool were difficult to meet. It had to be low-cost, so as to be available to house-needy rural workers, farmers, and other people of small means. It had to be light, portable, and simple to operate and to maintain. As mentioned in an earlier program, one of the stipulations was that it should be transportable by burro and repairable in any backwoods blacksmith shop.

The 140-pound Cinva Ram is such a portable, hand-operated press for making block and floor tile, and it is available for purchase in North America at $175, from Bellows-Valvain, 200 West Exchange Street, Akron, Ohio 44309. It consists of a metal mold in which damp, stabilized earth is compressed by a piston which is moved by a hand-operated lever mechanism. As a result of this ingenious lever system, a 70-pound manual pressure produces a compression force of 40,000 pounds, a ratio 20 times better than that produced by the Landcrete machine. The blocks are then extruded by a reverse action of the lever and removed to damp cure for one week. They should be air-cured an additional week before being laid.

Three men, doing all the work of processing, mixing, and molding should make 300 blocks a day. Production can be doubled by using a mechanical (cement) mixer and an earth-moving tractor. The only problem owner-builders seem to encounter when making Cinva Ram blocks is getting the correct amount of soil mix in the machine each

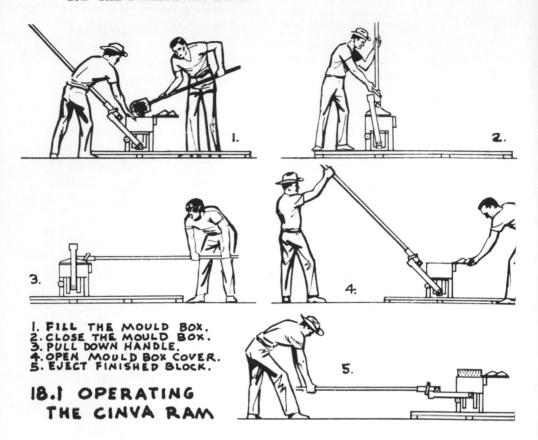

1. FILL THE MOULD BOX.
2. CLOSE THE MOULD BOX.
3. PULL DOWN HANDLE.
4. OPEN MOULD BOX COVER.
5. EJECT FINISHED BLOCK.

18.1 OPERATING
THE CINVA RAM

time. Fast, simple weighing equipment can be devised, but a combination of metal scoop and adjustable scraper seems to work exceedingly well.

The procedure for making Cinva block is very simple. First, a scoop of soil mix is placed in the mold box, and the cover is slid across. A 100-pound pressure is then applied to the lever arm as it swings over the top of the ram. The lever arm is, then, transferred to a person standing on the opposite side of the machine, who in turn applies the necessary pressure on the lever arm to eject the block. The block may be easily picked up without breaking if one applies a slight jiggling pressure to the block's width, not its length. The block is immediately removed to the curing area.

For stabilized blocks to gain strength, a moist curing period is essential, particularly if the blocks contain stabilizers of the cementing type. Unlike adobe blocks, which are sun-dried, pressed blocks need to be cured in a moist, shady environment. When feasible, there are advantages to erecting the roof structure of one's home before the earth walls are built and even before the blocks are produced. The roof can be either temporarily or permanently supported by well-braced, heavy, upright studding. Thus, this roof shelter, which is erected prior to the block-making, can be used to advantage as a curing space for the earth blocks as they are made as well as a shelter for the walls as they are erected. Also, the shade may be welcome to the builders for their personal comfort in hot weather and as a shelter for their tools and supplies.

Soils suitable for making pressed block are more limited than those for adobe block. For one thing, the clay to sand proportions are more critical. Soils should not contain more than 50 percent clay, including loam and silt, and not less than 50 percent sand, including gravel or other granular material. Experience indicates that the proportions are clay, 30 to 35 percent; sand, 65 to 70 percent.

Moisture content in the mix should be very carefully regulated. A wall's durability and its resistance to cracking depend upon the percentage of water content in the block, among other factors. The amount of shrinkage and cracking varies with the amount of moisture present, provided there is enough moisture to bond the soil particles. A simple test should be made to determine the moisture content of the soil to be used. First, sift a sample of this soil into a pan, and oven dry it. Then, place 8 pounds of the dried earth in a flower pot or a similar container having a hole in the bottom. Next, place the pot in a pan containing one pound of water. Through capillary attraction, the earth will absorb all of the water. The uniformly moistened soil will contain about 12 percent moisture by weight. This is the maximum percentage of moisture that should be allowed. Light, sandy soil of low collodial content should contain from 7 to 10 percent water.

A simple box test can be made to determine the optimum soil-cement ratio. The test determines the amount of cement that should be used by measuring the shrinkage in a soil sample containing no stabilizer. Inside dimensions of the box should be $1\frac{1}{2}$ inches deep, $1\frac{1}{2}$ inches wide, and 24 inches long. Moist earth, firmly packed in the box, is allowed to sun dry for three days. Shrinkage (contraction) is then measured by pushing the dried sample to one end of the box. If shrinkage is not over $\frac{1}{2}$ inch, the cement to soil ratio should be 1 part to 18 parts. If the shrinkage is between $\frac{1}{2}$ inch and 1 inch, the ratio should be 1 to 16. Shrinkage of 1 inch to $1\frac{1}{2}$ inches requires a ratio of 1 to 14.

Machine-compacted blocks have been found to be structurally superior to sun-dried blocks as well as to hand-compacted soil-cement varieties. So much so in fact that it is possible to reduce the outside bearing wall thickness to 6 inches, instead of building the usual 12-inch thickness. In Colombia, 6-inch earth blocks were used to construct buildings two stories in height. No more than 5 percent cement needs to be added to a machine-compacted block. One Colombian project, illustrated in the introduction, was built at a cash cost of $375.

At the Research Laboratory, Kansas State College, thorough tests were made on hand-tamped, soil-cement blocks. A block structurally superior to Bureau of Standards testing specimens was made, using a minimum of 7.5 percent cement to sandy loam which was 10 percent clay. This mix, slightly moistened, was tamped into a form and placed in a moist room for a two-week curing period. Identically proportioned, hand- and machine-compacted soil-cement blocks were, subsequently, made and structurally tested at Asawasi, Kumasi, Gold Coast. Piers 6 feet high and 18 inches square were constructed and loaded to failure. It was found that handmade blocks were crushed with weight at 43 pounds per square inch, while machinemade blocks crushed only at 121 pounds per square inch.

The final consideration for earth-block construction involves laying them up. Traditional, tried-and-true methods have changed little through the centuries. First, all wooden door frames are built and securely braced in position on the foundation. Then, a layer of block is casually placed around the building perimeter to establish relative block spacing for the remaining courses. When proper spacing has been determined by this trial method, the blocks are laid with mortar. The next step builds the corners. Corners provide a guide for laying the remaining blocks, so they should be laid level and should be correctly spaced. String lines are stretched between corners, even with the top of each brick layer. After a course of block is laid between corners, the string is moved to the following course. At window-sill height, wooden window frames are set in and securely braced in the same manner as the door frames. Blocks are then laid to lintel height. As mentioned in the previous chapter, a single reinforced concrete bond beam is cast on the top of the wall to function as a combination building tie, roof anchor, and lintel.

In their discussions on the use of earth material for homes, the Housing and Home Finance Agency claims that "it has been found that even the new techniques, which place emphasis upon careful soil selection, compaction, the addition of admixtures or a combination of any or all of these to increase resistance to water, impact, and erosion, need not be beyond the capabilities and ultimate needs of the average low-income people." This opinion is sound, and there is, hopefully, sufficient information in the following two chapters to qualify an inexperienced person to build earth blocks and to lay those blocks in a wall. But technical qualification is actually only a minor part of earth-home construction. None of the world-wide government agencies dedicated to promoting earth-wall construction show the same concern for earth-wall design. The rectangular-shaped, proto-type earth-wall house could just as well be built of straight-edged lumber.

18.2 EARTH-WALL DWELLING

Any experienced, observant earth-home builder is aware of numerous design improvements that should be made on the average earth-wall dwelling. Building corners, for instance, can become an annoyance. All jogs and unnecessary angle changes should be reduced. Better yet, eliminate corners altogether. Wall sections can be designed to intersect at openings. Openings, too, require inordinate amounts of labor. They should be grouped so that larger sections of wall panel remain uninterrupted by the piercings of occasional windows and doors. There is little advantage to building partition walls of earth block. It is the house perimeter that gains most from these highly insulating, fireproof earth walls.

Earth homes should maintain a low profile. They should certainly be built no more than two levels high. Block-laying labor efficiency is reduced dramatically as wall height increases. Where structural wall problems are encountered, as in earthquake or high-wind areas, consider curving the wall panels. Earth blocks are a plastic material before setting up and can be laid in a reasonable curvilinear form. A curved wall offers significantly more sheer strength and rigidity than a straight wall.

19 RAMMED EARTH

19.1 KIRKHAM RESIDENCE

The building illustrated above is, at first glance, quite ordinary and unpretentious. In plan, it has the typical five-rooms-and-bath arrangement. What gives this house distinction, however, is the fact that it was built as an experiment, using low-cost materials. Civil Engineering Professor John Kirkham constructed the building in 1940 at Stillwater, Oklahoma. The work was done by unskilled labor, and the total cost of the house when completed, including plumbing, lighting, and all finishing, was $887.80. This astonishingly low construction cost was made possible in part by the use of tamped earth in the walls, floors, and roof. Thin layers of reinforced concrete, $1\frac{5}{8}$ inches thick, were poured to support the flat earthen roof and to surface the tamped-earth floor.

Rammed-earth construction is older than history and at various times has been used in almost every country in the world. The evidence of centuries has conclusively proven rammed earth buildings to be sufficiently strong and lasting. Besides their cool-in-summer and warm-in-winter thermal characteristics, earth-wall buildings, in general, are dry, fireproof, rot-and-termite proof, and soundproof.

Rammed earth is a method of construction whereby continuous walls may be built by tamping moist soil into sturdy wooden or metal forms. When a short section of wall is completed, the forms are moved upward or sideways, and the process is repeated. Tamping is done with either hand or pneumatic tampers. Whichever method is used, as each layer of earth is placed in the form, the tamping should be thorough. Only when the noise from the tamping tool changes from a dull thud to a clear, ringing sound, can the wall section be considered completed.

Experimenters at South Dakota State College discovered that the compression strength of a rammed-earth wall can vary from 93 to 393 pounds per square inch, depending upon the intensity of the ramming stroke. In order to achieve maximum strength, the rammer should be raised about 12 inches and the worker should apply all the force possible with each stroke. The heavier the tamper, the faster the earth becomes compacted, yet even the heartiest laborer tires when the tamper weight is over 25 pounds. In general, the tamper should weigh 2 pounds for every square inch of tamper force. The most common size, for instance, has a 3-inch-square face and weighs 18 pounds.

Much labor and time are involved in tamping earth in a form. Both can be reduced one-half by using an air tamper. Use of this tool becomes significant when one considers that it takes a workman an hour to properly tamp 4 cubic feet of wall by hand. A long-stroke

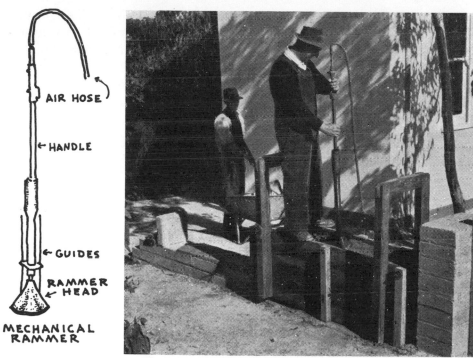

AIR HOSE

← HANDLE

← GUIDES

RAMMER
← HEAD

MECHANICAL
RAMMER

19.2 WALL BUILDING   AUSTRALIAN BUILDING STATION PHOTO

machine of moderate feed and light weight, 25 to 30 pounds, is ideal. This machine is similar to the one used for tamping back fill into trenches and for tamping sand into molds in heavy foundry work. Powerful strokes will result from using an air compressor capable of supplying 70 pounds of air per minute. An air compressor with a free-air delivery of 25 cubic feet per minute will operate one tamper.

In one sense, a rammed-earth wall is a large, pressed earth block, which is cast in place *in situ*. The construction of the form that holds this block has a lot to do with the success or failure of one's wall building. Traditional wall forms have been cumbersome and heavy, requiring much adjustment, bracing, and alignment. Just having to use these forms is more than enough to discourage one from building a monolithic earth wall. But in recent years a number of improvements in form design have been made throughout the world. At the Texas Agricultural and Mechanical College, lighter weight plywood was substituted for the heavy planking usually employed. The Commonwealth Experimental Building Station in Australia developed a roller-supported form with detachable wooden clamps. They found that the old-style, demountable form required $1\frac{1}{2}$ hours for three men to dismantle, reset, and plumb, whereas the moving and plumbing of their roller-supported form averaged 8 minutes and required but one man for the entire operation.

Some years ago, contractors Dan and John Magdiel built more than their share of rammed-earth homes in the Southwest. They developed and, subsequently, patented a form for their own use and for manufacture before their deaths in 1966. The Magdiel wall form is entirely metal and is well braced. After a form-tamped section of earth is thoroughly tamped, one merely pulls up on a release lever to loosen the sides of the form. It is then moved ahead, clamped into place and tamping is resumed. Traditional systems of wall forming require a corner section as well as a straight-run section. Magdiel eliminates the bother and the expense of a corner form by simply tamping one straight section at right angles to another. A United States Department of Agriculture bulletin on rammed-earth construction suggests notching the end of the form so that it will pass, unobstructed, over the adjoining wall. The corner is then made by right-angled, alternating courses. Design of the rammed-earth form illustrated in Figure 19.3 has evolved from my many years of experience building monolithic earth walls. This form was devised to be adjustable for any corner angle and to retain the rolling feature of the Australian form, which is so desirable for straight-wall forming. This form can be braced with detachable wooden C-clamps, instead of with bolts. The legs of the clamps support the form's sides by the cantilever principle. The form's sides are, thus, externally braced and there are no bolts to remove and to replace before tamping the next section. The form is light in weight and can be easily handled by one man. It is also quite inexpensive and simple to construct. To maintain a plumb and straight wall, the form must be leveled each time it is set up. Door and window frames should be made the same thickness as the wall, 12 to 16 inches, and be placed in proper position in the form before tamping. Prepared earth is shoveled into the form to a depth of 4 inches and is firmly tamped. Sides are tamped first, working toward the center. The 4-inch layer will compact to about 2 inches. When tamping is completed,

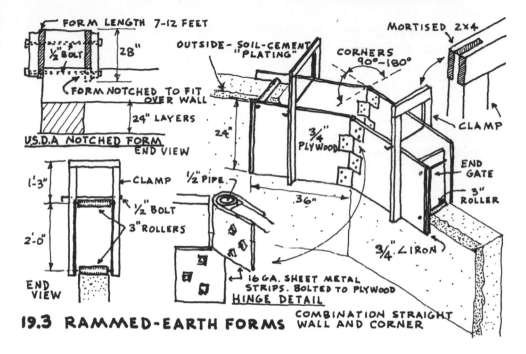

FORM LENGTH 7-12 FEET

½" BOLT   28"

FORM NOTCHED TO FIT OVER WALL

24" LAYERS

U.S.D.A NOTCHED FORM
END VIEW

OUTSIDE- SOIL-CEMENT "PLATING"

MORTISED 2×4

CORNERS 90°-180°

CLAMP

24"   ¾" PLYWOOD

1'-3'   CLAMP   ½" PIPE

½" BOLT

3" ROLLERS

2'-0"

.36"

END VIEW

16 GA. SHEET METAL STRIPS. BOLTED TO PLYWOOD
HINGE DETAIL

END GATE

3" ROLLER

¾" ∠ IRON

19.3 RAMMED-EARTH FORMS   COMBINATION STRAIGHT WALL AND CORNER

the form is moved along to the next section, one end being lapped at least 4 inches over the work just completed.

For about $100, the outside walls of an earth home can be stabilized by adding common cement to the mixture. As a result of adding 15 percent cement to a very fine, sandy-loam soil, which is 74 percent sand, the strength of the wall increased from 302 to 1,150 pounds per square inch. A stabilized wall is affected less by moisture penetration and absorption, and it is more resistant to other weathering agents. Cement is most economical in soils containing 50 percent or more sand. With this proportion, about 15 percent cement should be used, whereas, with the more ideal proportion of 70 percent sand, only 8 percent cement is needed to achieve the same result. With a 60 percent sand mix, about 12 percent cement should be added.

In addition to having structural superiority, a soil-cement wall will shrink less than one-half the amount of a plain rammed-earth wall, $\frac{4}{10}$ inch in 12 feet as against 1 inch in 12 feet. The cost of the additional cement in the mixture can be reduced two-thirds by plating the outside, weather-exposed wall surface with soil-cement, and by filling the remaining space with regular rammed-earth soil. It is a simple operation. A shovelful of the soil-cement is thrust against the outside wall of the form, and three or four shovelfuls of earth without the cement are placed against the inside wall of the form. This four- or five-inch-thick layer of earth is then tamped in place.

A number of other stabilizing agents besides cement can be used in a rammed-earth wall. An owner-builder should not hesitate to experiment with various locally available stabilizers. Cement seems to work best with sandy soils. Lime, either slaked or unslaked, makes one of the best stabilizers for clay soils. It is not as expensive as cement and it is certainly more available. Lime also mixes well with fly ash, the fine dust given off during the burning of coal. Wood ashes, too, have been used to stabilize soils. Fine white ash from fully burned hardwood seems to work best.

Rammed-earth walls are often left untreated, some surviving hundreds of years of exposure. Ralph Patty wrote in *Age Strength Relationship for Rammed Earth,*[*] that, with age, earth walls increase in strength and in erosion resistance. Tests of three widely varying soils indicated an average strength increase of 34 percent with one year's age, and 45 percent with two year's age, as compared with a wall's strength at six month's age.

The Magdiel brothers developed an interesting method for waterproofing rammed-earth walls. Single strands of wire were placed across each layer of wall section as it was built. Then, when the wall was completed, they attached chicken wire to both sides of the wall, secured by the tie wires previously set during tamping. The wall was then plastered inside and stuccoed outside.

For the sake of economy, it would be well to even waive use of the wire mesh for plaster backing, if possible. For a long time, I wondered how it was that Africans achieved such excellent plastering results without wire-mesh backing, whereas in this country many precautionary measures (that is, mesh, masonry ties, chipping, and so forth) must be observed to insure a good bond. The answer to that uncertainty is that in Africa the proportion of cement to sand is in the nature of 1:12, plus a 5 to 10 percent mixture of lime to the

*Ralph Patty, *Age Strength Relationship for Rammed Earth,* Engineering News Record, vol. 117, no. 2 (1936), p. 44.

plaster, whereas we are accustomed to using a 1:3 proportion of cement to sand and about 10 pounds of hydrated lime to each bag of cement used. Apparently, the coating of a strong material, cement and lime, over a weaker material, earth, is basically in error. Differences in expansion and contraction are apt to cause cracking, breaking of the bond, and ultimate failure. The 4,000 rammed-earth homes built in India (and mentioned in a previous chapter) were all successfully protected with a weak stucco composed of a 1:15 cement-sand mixture applied over a cement-wash undercoating bond of 1:3 cement-water.

Possibly the most ancient of all wall finishes is Dagga Mud plaster. Quality Dagga plaster contains enough fine sand to let the plaster dry without checking. The 3 parts sand to 1 part clay should be screened through an ordinary fly screen. Sometimes lime (1 to 8 percent), cement (10 percent), or even asphalt emulsion (1 gallon to 100 pounds of dry soil) is added to the Dagga. The wall is first dampened, then a primer coat of plaster is applied. It must be kept damp during the early curing period. Two thin coats are usually applied.

In Southern Rhodesia, rammed-earth walls are often sprayed with a bitumastic emulsion which, when tacky, is harled with sharp, clean sand thrown against it. A cement wash is then applied to the wall. It has been found that a coat of asphalt-based aluminum paint will bond very well to asphalt-stabilized earth walls. Aluminum flakes in the paint lie flat and overlap, preventing any leaching through of the asphalt in the wall. A finish of oil paint can then be applied. Linseed oil is also a very successful and widely used protective covering for earth walls. Two or three brush coats should be applied, followed by an interior finish of one or two coats of household paint, if desired.

Dusting of interior earth walls can be prevented by application of a homemade waterproof glue. This waterproofing application is made of 6 parts cottage cheese and 1 part quicklime, with sufficient water to make it flow smoothly. The use of cottage cheese for glue sizing may sound a bit odd, but it must be remembered that most of the original research on rammed-earth finishes, especially those done by the Engineering Department at University of South Dakota, had as its purpose inexpensive construction for farmers. This research group even developed a buttermilk paint: $4\frac{1}{2}$ pounds of white cement to 1 gallon of buttermilk. Two coats should be applied over a glue-sizing primer coat.

Experimenters at the University of Colorado had the grain farmer in mind when they developed a flour and water covering. Thirty pounds of flour and 50 gallons of water are cooked to a creamy consistency and are then added to a mixture of the same soil that is used in the walls. Then more water is added to thin it for application.

Professor Kirkham's bulletin describes the type of brush stucco he developed for both inside and outside treatment of his $887.80 house. His two-coat stucco is very cheap, does not crack, and is absolutely waterproof. It should be applied with a stiff brush:

> Place a cake of laundry soap (stearic acid soap) in a three-gallon bucket and fill the bucket with cool water. Then, take the cake of soap in your hands and rub the soap, as you wash your hands, until the water becomes quite soapy. This soapy water is used instead of plain water when mixing the stucco. Next, place four-pints of Portland cement (4 pounds) into another three-gallon bucket and to this add four pints of fine sand. Then, using a small wooden paddle, stir until the cement and sand are well mixed. Add the soapy water to this dry mix and keep stirring until the mixture becomes a stiff paste which may be applied directly to the wall. After the paste has been on the wall for 24 hours, the surface is painted (washed) with a copper-sulphate solution which is obtained by adding a quarter of a pound of copper sulphate crystals to three gallons of water.

20 STONE MASONRY

Next to earth, stone is the most natural of all native building materials. It is reasonable to suppose that in areas where building stone existed, man's earliest dwellings resembled mere piles of rock. Then, somewhat later, he probably found that the drudgery of handling so much material could be eliminated by laying one stone upon another. We know to this day that the strength of a stone dry wall depends entirely upon the "firmness of seat," that is, the fit of one stone on top of another. Then man's discovery of natural cements made it possible for him to build wet walls, walls that can be at once thinner, higher, and stronger.

It is a mistake to attempt a minute analysis of the advantages and disadvantages of stone-masonry construction. For all the advantages of any material, one can find disadvantages. A stone wall is, for instance, relatively fireproof but requires some form of waterproofing. A stone wall is strong and has a long life, but its strength depends entirely upon the workmanship in laying it and upon varying mortar proportions. A stone wall requires little maintenance, but there is a high initial labor cost for its construction. The comparatively small size of individual stones aids in their handling, but the wall itself has great weight. A stone wall is relatively soundproof, but additional insulation against heat transmission is required. A stone wall can be built to hold heavy roof loads, but the wall can scarcely resist bending or tensile forces. If you respond to a challenging material, if you want a wall that is unique and beautiful, with great potential for individual expression, and if stone for this wall is locally available at little cost above hauling, then you may well consider stone-masonry construction.

Like earth construction, which has been used for centuries in building walls, floors, and roofs, stone is readily available at little or no money cost. It can be gathered, usually free for the hauling, from any stream bed, from abandoned mines and quarries and from open fields and embankment cuts. There is hardly a region in the country that does not contain a substantial resource of building stone.

Maps and aerial photographs of one's region are generally available and can be employed to advantage in locating building stone. Agricultural soil maps are revealing and thorough. Geologic maps indicate existing pit and quarry sites, as well as the type and the structure of the stone in those pits and quarries. U.S. Coast and geodetic survey maps cover nearly every section of the country. They are especially helpful in locating abandoned ore mines. Tailings from mines are among the best sources of building stone. From aerial photos, one can locate such stone-laden features as excavations, outcroppings, cliffs, abandoned railroad and road cuts, and natural stream beds.

With such widespread availability, one asks, "Why is building with stone so rarely exploited by home builders?" The answer to this query is that building with stone is similar to building with earth. There are large time-and-labor factors involved in gathering and placing the material in a wall. But the average owner-builder's time and labor resource customarily outweighs his capital resource, so this cannot always be considered a serious handicap.

Perhaps a more pertinent answer to why building stone is seldom used lies in the fact that stone-masonry technology, more than any of the other building-trade skills, has been traditionally clothed in secrecy. Carl Schmidt, in his little book *Cobblestone Architecture*, illustrates this point:

> Several very old men, who as little boys saw cobblestone masons
> at work, readily recall the jealousies among the masons. Whenever
> a visitor appeared while they were working, they would stop

work, hide their tools and do something else until the visitor went on his way. The fact that these men succeeded very well in keeping their own methods a secret, explains the different mannerisms found in the method of laying up the walls.

Through the centuries, stone masons also have succeeded in maintaining a respectable, highly paid, and somewhat apostolic status in the building industry. Their trade secrets are maintained to this day and include such important items as an intimate knowledge of the nature of stone itself, the correct mortar proportions and the use of auxiliary materials, the proper selection of tools and organization of work procedure, and, finally, an aesthetic awareness of the stone-in-place, that is, the total effect and composition of the finished wall. Intensive research on stone masonry reveals that no pertinent literature on the subject of stone masonry is available to the unskilled owner-builder. Stone masons maintain their closed shop. In this chapter, I will make an attempt to close this enigmatic gap.

With fear of oversimplifying the stone-masonry skill, it should be stated that the foremost prerequisite of any mason worth his mortar is an intimate, nearly intuitive, knowledge of rock. Pick up a rock. Where the inexperienced stone worker observes color, weight, and form, the experienced stone mason notices bedding, seams, rift, and grain. He first visualizes the rock in place, laid on its natural bedding. Bedding refers to a rock's horizontal stratification as a result of deposition in sedimentary layers under water. It is seen in the granular changes in color and in texture in the horizontal joints of sedimentary rock in which it is most prevalent. Seams are generally vertical to the rock surface. Seams are regular in their direction in limestone and they are irregular in granite. They occur in rock as a result of compression and tension stresses in the earth. The direction for greatest ease in splitting rock is called the rift. It may be parallel to the seam. A second, minor direction for splitting is called the grain. Only the most experienced mason can detect a rock's grain direction.

Stone has much the same physical characteristics as wood. Masons talk of the sap in freshly quarried stone. After exposure for some time this sap crystallizes, and the stone becomes harder and better able to resist weathering by the elements. The rift direction of a stone is an important thing to ascertain when building an exposed wall. Limestone, shale, and sandstone have stratified rift, a property that makes them easy to split and to chip, but this property also makes them vulnerable to cracking in winter, when they absorb moisture which turns to ice during a freezing spell. In most hand-laid work one's first tendency is to place the stone with the rift lying horizontally and running from the outer surface of the wall to the inside of the wall. But then capillary action draws water to the inside of the wall via natural splits and seams. It may also draw water in-between stone and mortar. A few coatings of masonry sealer will help this problem, but the best practice is to place the stone on its splitting surface, with the rift in vertical position.

A number of stone preservatives are available, designed to protect rock from the aforementioned frost and moisture penetration hazards. A waterproofing agent prevents the penetration of moisture, but moisture that does gain access into the wall is not permitted to escape through the same waterproofing. This is a troublesome situation. The wall should breathe, whatever material is used. Moreover, the outer waterproofing layer is a thin skin that differs in physical properties from the underlying material. This difference causes certain stresses to be set up which inevitably force the outside skin to flake off.

Several simplified systems of rock identification have been devised to assist the mason in his choice of building stone. Rock classification can be physical, differentiating between stratified and unstratified rock, or it can be a chemical classification, dividing rock into its siliceous (sandy), agrillaceous (clay), or calcareous (lime) composition. The classical classification of rock, however, is based upon geological origin; that is, its igneous, sedimentary, and metamorphic origins. A composite classification system of the more common building stones, along with their significant construction properties, is presented in Table 20.1.

The dearth of books on the subject of stone masonry may be a continuance of the closed-shop stone-masonry conspiracy mentioned earlier. In any event, there are no contemporary manuals on laying up building stone. The Audel reference text on masonry is typical of what is currently available for beginning masons. The stone-masonry techniques and the tools discussed in this text date back to antiquity. Only the correct hammer and chisel are identified, and the manner of squaring huge marble building blocks is inconsequentially described.

TABLE 20.1

CONSTRUCTION PROPERTIES OF BUILDING ROCK

Geologic origin	Physical type	Rock name	Mechanical strength	Durability	Surface character	Presence of impurities
Igneous (formed from molten material)	Intrusive coarse grained	Granite Diorite	Good Good	Good Good	Good Good	Possible Possible
	Extrusive fine grained	Basalt Obsidian	Good Good	Good Good	Good Good	Seldom Possible
Sedimentary (sediments deposited by wind and in water)	Calcareous calcite	Dolomite Limestone	Good Good	Fair Fair	Good Good	Possible Possible
	Siliceous silica	Shale Sandstone Chert Conglomer	Poor Fair Good Fair	Poor Fair Poor Fair	Good Good Fair Good	Possible Seldom Likely Possible
Metamorphic (prolonged heat and/or pressure)	Foliated parallel layered	Slate Schist	Good Good	Good Good	Poor Good	Seldom Seldom
	Nonfoliated	Quartzite Marble Serpentine	Good Fair Fair	Good Good Fair	Good Good Poor	Seldom Possible Possible

A number of unlikely research sources have been used to compile this chapter on stone masonry, but, primarily, my actual stone-laying experience over the past fifteen years forms the nitty-gritty of what follows.

The rock classification system illustrated above in Table 20.1 can prove of only general value to the owner-builder mason. Let's have a closer look at choosing rock and at building with this natural resource. Accessibility of the rock must be one of the prime criteria. An expensive quarrying or hauling operation can be a deterrent sufficient to dissuade one from using this material in a building. In some instances, a particularly hard rock is called for, as in floors and in steps. Rock with cleavage, a splitting quality, is generally a more valuable characteristic than a blocklike monolithic quality.

Of course, we desire to build a durable wall, one that will withstand rain, wind, frost, heat, and fire. Building-stone life ranges from 10 to 200 years. Frost damage is common to softer, porous rock. Again, if rock is not laid on its natural bed face, frost action will tend to separate the layers, as illustrated in Figure 20.2. Another important rule is that the strength of the mortar should be equal to the strength of the rock. An excessively rich mortar is more pervious to frost than is a weaker mortar, because shrinkage cracking occurs in rich mortar. The most vulnerable part of the wall to moisture penetration is the mortar joint.

Granites are least affected by weathering. Limestone and sandstone are, on the other hand, the most affected by weathering. They are commonly destroyed by surface erosion, from sea salts, for instance, and from atmospheric pollution. Rain will leach the cementitious materials found in some sandstone to the surface, where they become brittle, weak, and finally flake off.

One may reason that strength should be the foremost requisite of rock for building purposes. However, rock that is sound and suitable in other respects is almost invariably strong enough for use in a wall. Recent tests at the U.S. Bureau of Standards on samples of Montana quartzite indicated a compressive strength of 63,000 pounds per square inch, a rather typical rock strength. A structure of such material would have to be over 10 miles high before failure would occur from a crushing of the lower courses. Another good example of structural rock strength is illustrated in the 555-foot-high Washington Monument. Pressure at the base course is 700 pounds per square inch, but marble will sustain a crushing load of 25,000 pounds per square inch.

The appearance of your dwelling should be considered when choosing a building stone for your home. Each rock has its own unique color, and rock of different colors can be mixed in a wall. Every rock also has its own unique luster, whether it is vitreous, pearly, resinous, dull, or metallic. Rock containing much iron should be avoided, since stains caused by oxidation of the iron will discolor the mortar.

Some rock can be worked better than others. Angular, square-edged quarried rock lays up better than roundish cobblestone boulders. These last are sometimes called rolling stones, because they are loosened and weathered from the parent ledge by natural processes.

Workability depends as much upon the correct mortar mix as it does upon the type of rock laid. A proper mortar is weather resistant

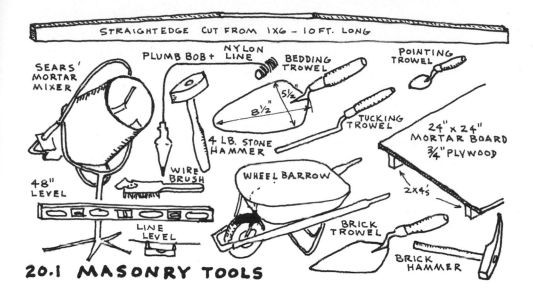

20.1 MASONRY TOOLS

and has adequate bond strength and compressive strength. The proportions of sand, cement, fireclay, and, especially, water must be controlled within a narrow margin. The optimum proportion is 12 shovels of clean, washed concrete sand, 3 shovels of common cement, and 1 shovel of fireclay.

The actual process of laying stone consists, first, of spreading a uniform layer of mortar and, then, of forcing stone into it, a process called bedding the stone. The mortar should be stiff enough to support the stone without letting it touch the stone underneath.

A bedding trowel is used by stone masons for spreading large mortar beds. Unlike a brick mason's sharp-pointed trowel, the bedding trowel has a rounded tip. Two sizes are commonly used. The 2½-cubic-foot, detachable-steel-drum, concrete-mortar mixer sold by Sears is entirely sufficient for either small or large masonry jobs.

After a course, or layer, of stone is laid, the wall behind the facing stone must be carried up to give support to the face. This is termed backing and usually consists of a cheaper class of masonry or of poured concrete, bonded directly to the face. Bondstones act as ties, bridging face rock to its backing. Metal strips, masonry ties, are commonly used to secure the face to the backing.

The simplest, fastest, and, in all respects, the neatest type of stone-masonry pattern for the owner-builder to work is called cyclopean masonry. Various sizes and shapes of stone are used, with no respect to regular course. Deep-cut joints, spaces between stones, look best. A ½-inch-wide tucking trowel is used for this purpose. Master stone masons can rightfully be proud of their time-consuming, varicose vein mortar joints, but the final result does not compensate the effort required to form these detailed, stylized joints.

There are several design features of cyclopean masonry that are essential to a wall that reads well. First of all, it is necessary to break or stagger the intersections of the mortar joints. Then, too, rock sizes should be well proportioned and graded from the small spalls, which are rocks that are filled into spaces too small for regular-sized rocks, to the larger-size stones, which are proportional in size to the size of the wall. Triangular-shaped stone, or long sliverlike specimens, should be placed to give a directional vitality to the wall. A triangular stone, with its apex pointing down, gives a more dynamic impression than if it is placed with its point upward. A common error that most amateur stone masons make is to congregate large-size rock near the base course and to finish the upper courses with progressively smaller and smaller sizes of rock. This looks as if the builder ran out of good material as his work progressed. Corners are always set first in wall work, although edges are laid first in flatwork, such as in slate flooring, and so on.

The stone wall panel illustrated here is an example of better-than-

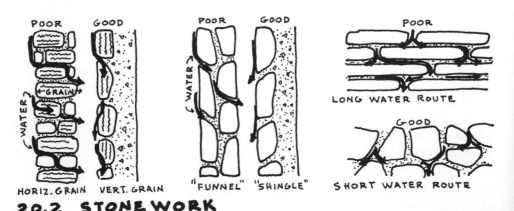

POOR   GOOD   POOR   GOOD   POOR

←GRAIN→

←WATER→   ←WATER→   LONG WATER ROUTE

GOOD

HORIZ. GRAIN   VERT. GRAIN   "FUNNEL" "SHINGLE"   SHORT WATER ROUTE

20.2 STONEWORK

average masonry. The rock forms are natural and restful, and rock sizes are pleasingly proportioned to the total size of the panel. Triangular, square, and random shapes are thoughtfully distributed to create a dynamic, readable composition. Deeply recessed joints assist the eye in its movement and in its regrouping experience.

The most obvious regrouping consists of rocks 3, 11, 12, 10, 27, 18, 26, and 29. A readable directional quality is achieved in the sequence

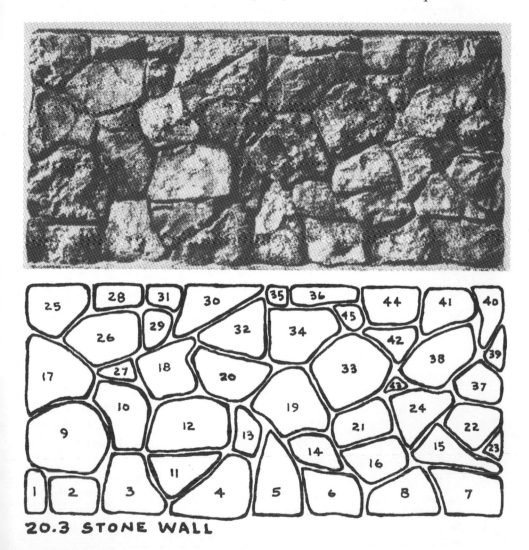

20.3 STONE WALL

of these rocks themselves without lining up joints, as in brick or block work. Notice how rock 34 breaks the joint line between 19 and 33, 14 and 21, 6 and 16 and 8. Vitality is also achieved by strategically placing triangular forms such as rocks 24, 11, and 30. The downward-pointed apex of these rocks adds a dynamic, unbalanced aspect to the composition.

A final feature that qualifies this panel for professional status is the thoughtful placement of base, corner, and top rock courses. Top corner rock 25, for instance, is more massive than bottom corner rock 1. Base rocks 1 and 2 are overpowered by corner rock 9. Top rock 30 complements its lower neighbor rock 32, thereby creating a regrouping which consists of rocks 30, 32, 20, and part of 18.

A few detractive criticisms of this panel may also be in order. Rock 43 is the only spall, or fragment, used in this panel even though places exist for at least a few more such rocks between rocks 8 and 16 and between rocks 13 and 12. Notice how beautifully spall 43 integrates neighboring rocks 33, 38, 24, and 21. Corner rock 37 should never have been used. Its top slope makes it difficult for the builder to set the next corner rock 39. Top corner rock 40 adds, further, to this conflict. Its effect is to wedge out rock 39 at the top, while at the same time it appears to be slipping from its bed. The left-hand side of this panel has much more stability and grace than the right-hand side.

The sequence of rock laying is indicated numerically. Notice that 1 begins at the left-hand corner, and that work is carried to the right. Corners are always set first and then interior spaces are filled in. Generally, larger rocks are set first, with smaller ones filled in around them. It is simpler to fit smaller rocks around large ones than it is to find a place for a large rock.

Large rock 26, for example, is bedded on rocks 17 and 18, and small rock 27 is set after the cavity has been fully defined. Top rock 30 is temporarily propped into position so that its top is level with the top of the wall. Filler rock 32 is then found to fit the cavity. Small rocks, especially spalls like 43, are always set after the larger rocks are in place.

The main element in any type of hand-laid stone masonry is simply time. Under average conditions a mason and his helper can place about 2 tons or 60 cubic feet of wall in one day. Including cost of labor and materials, the contract price for stone work starts at $2 a cubic foot. Consequently, our patent office contains hundreds of applications for devices that purport to reduce this costly time element.

In 1920, a New York architect, Ernest Flagg, developed the first truly successful labor-saving device for the construction of rock walls. His mosaic-rubble system of forming walls has the advantage of being totally integral, as no external bracing is required. The form is attached to and rests on 4-×-4-inch sleepers, which are embedded, four feet apart, in the concrete foundation wall, while it is being poured. Then, on each end of the 2-foot-long sleepers, a 4-×-4-inch upright is fastened, its length equal to the desired wall height. A movable 12-foot-long alignment truss is employed at the upper limits of these vertical uprights. Between the foundation sleepers and the alignment truss, the rock wall is laid. Movable 2-×-6-inch or 2-×-8-inch planks rest against the inner sides of the uprights. When one layer of the rock and concrete reaches its initial set, these planks slide upward and are held in place by toggle pins. In practice, the rock is merely set against the outside plank, and the rest of the space is poured with a weak

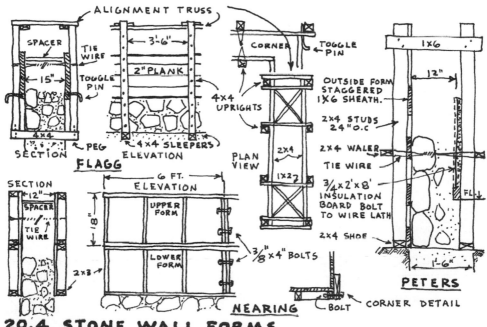

20.4 STONE WALL FORMS

(1 cement, 5 sand, 10 gravel) concrete mixture. When the forms are removed, the outside wall may be pointed, whereby cement mortar is troweled between the rock joints, or it can be left untreated for a "rough" effect.

Following Flagg's original scheme, another Eastern architect, Frazier Peters, built a number of mosaic-rubble houses. Later he devised his own personal system of wall forming, which involves erecting a complete wooden shell on all four walls and placing the rock and mortar against this. Peters' system appears unnecessarily involved, although he patented a few rock-wall building ideas which are well worth consideration.

In the 1930s, Helen and Scott Nearing started to build rock structures on their Vermont homestead. Their program was an ambitious one, and so they naturally considered in detail many of the different forming methods suitable to amateur owner-builder construction. An adaptation of Flagg's mosaic-rubble form was employed by the Nearings in the eventual construction of nine rock buildings. First, they assembled a large quantity of 18-inch-high plywood forms of varying lengths. When set in position on a wall, each pair of forms was fastened together with wires, a procedure suggested by Flagg. After a first set of forms was filled with rock and concrete, using a proportion of 1 part cement, 3 parts sand, and 6 parts gravel, a second set of forms was placed on top and filled. When the concrete in the upper form reached its initial set, the wires in the lower form were cut and that form was placed on top of the upper one and fastened with new wires. This climbing-form method of rock-wall construction proved to be an ingenious improvement over Flagg's plank-and-stud system.

Movable wall-building machines, like the Magdiel form, can also be used successfully for building rubble walls. Labor costs can be halved with the use of this form, and amateur masons can produce walls of professional quality with the help of this simple mechanism. Another type of movable form can be made out of sheathing, braced from the inside. This panel should be large enough to take care of all the stone that can be laid in one day's work. The following morning, after the wall from the previous day's work has set, the form is moved to a new position, and the rock laying is continued. Since the rock laying is done from the outside of the form, the inside can be braced to the floor or to window and door frames.

Various methods for insulating stone walls have been tried. The standard practice for insulating masonry walls uses furring strips set vertically flush in the masonry. These members serve as light nailers for a second set of nailer strips, which is horizontally fastened to the first set of vertical furring strips. Insulation board is secured to these horizontal nailers. The embedded strips are apt to dryrot with age, and, unless a great deal of patience is exercised in setting the strips straight and true, the wall is certain to be uneven.

Perhaps a better method for providing insulation against the inside of a poured, rubble wall is to cast the wall against wood-stud framework. This method was used in building the Fryer House, designed by the author and illustrated below.

The author is partial to any construction that is simple enough to be understood and to be implemented by anyone having average intelligence and work ability. The complicated technique of masonry construction seems to require a few basic aids before the average man can master it. Flagg, Peters, the Nearings, and the Magdiels all suggest methods that lessen the need for high level masonry skills. All these

20.5 CAST STONE   PHOTO: RICHARD FRYER

systems work well and contribute greatly to the advancement of owner-builder construction.

Design subtleties in stone laying can be incorporated, which add immensely to the interest of the building as well as enhance the value of the masonry work. In a building, a harmonious interplay of stone, wood, and glass is always to be sought. Stone should contrast with these other building elements as well as with the native surroundings. On a sloping site, for instance, a massive stone foundation wall binds the building to the sloping terrain. It links the natural landscape to the formal discipline of the building.

Success in building with rock is not easily come by. But no material blends as well as stone does with the natural environment or better reveals the personal artistry of the builder.

21 MASONRY BLOCK AND BRICK

If the reader is not already aware of the basic organization of the materials section of this book, it will soon become obvious that materials are here classified into three general groups, depending upon the degree to which the material is modified. Synthetic materials, like plastics and paper-core panels, are greatly modified; manufactured materials, like plywood and concrete, are modified to a lesser extent; natural materials, like stone, earth, and timber, are only slightly modified.

Obviously, the less processing required of a material, the less it will cost in the finished wall. Modern building technology has produced some truly remarkable wall materials, but, without exception, the saving made in added insulation, speed of erection, durability of finish, and so on, is forfeited by the higher cost of the material itself. Inasmuch as the average owner-builder has a labor resource far in excess of his money resource, it behooves us to consider in great detail the less-processed, natural building materials.

From 100 to 150 million Asian families live in substandard, crowded, unsanitary housing. This estimate comes from a mission of the United Nations on tropical housing. The basic problem of housing for Asia's millions was presented at the opening session of the 1953 United Nations Seminar on Housing and Community Development. American dwellings are poorly conceived and poorly executed, but they are mansions by comparison to those of Asians. It has long been taken for granted in America that the world's masses, the native peoples of Asia, Africa, and Latin America, will remain miserably housed. It is taken for granted no longer, especially by the insurgent masses themselves.

Research on housing for underdeveloped areas has begun. The National Building Institute, under the auspices of the South African Council for Scientific and Industrial Research, began a research project in 1954 on low-cost housing for the urban Bantu. Its final report, "Research Studies on the Costs of Urban Bantu Housing," a thorough piece of work detailing the social, economic, and technical factors of housing, establishes minimum standards for building performance.

Curiously, this capable research staff chose the familiar brick as the most satisfactory building material to meet basic housing requirements. The reasons for this choice can be appreciated when the nature of the material is understood, and when one observes a master brick mason at work. For one thing, the size of burnt-clay brick has remained nearly constant since the manufacture of these elementary building units evolved from the baking of clay pottery in ancient Egypt. One advantage of small building units is their adaptability to practically any design. The size and weight of a brick is, moreover, perfectly scaled to human size. Also, a work rhythm develops as a master brick mason establishes his balance, picking up a brick with one hand and a trowel of mortar with the other hand.

The brick-on-brick procedure of wall construction, however, is far behind some of the monolithic wall-forming methods. Trade organizations, like the Portland Cement Association, the Structural Clay Products Institute, and the National Concrete Masonry Association, were actually the first to anticipate the waning use of individual masonry units. These organizations have spent millions of dollars developing improved masonry units, tools, and equipment for speeding brick and block construction. As is usually the case in such highly organized, private projects, the individual owner-builder has seen little of this invaluable research.

The ancient, slow, brick-on-brick method of construction persists.

But, where the professional brick layer is too routinized to experiment with an unorthodox masonry unit, the amateur owner-builder, with his open mind, can profit by the experimental approach. The purpose of this chapter is to familiarize the owner-builder with the standard practices of brick and block laying and with some of the new departures in wall construction inside and outside of the masonry-block industry.

Masonry units have at least one structural advantage over monolithic construction, for example, poured concrete. As small units they are, in effect, preshrunk. Less shrinkage cracking takes place, and, therefore, less reinforcement is customarily required. As with any material, however, the quality of the workmanship plays a decisive role in the structural value of the building. The strength and resistance of clay-brick walls to penetration by rain depends more upon the completeness of the bond between the mortar and the brick than upon any other single factor. According to laboratory tests at the National Bureau of Standards, partially filled mortar joints result in leaking walls and a reduction in the strength of the masonry in the range of from 50 to 60 percent. All joints in both the backing and the facing should be completely filled with mortar as the brick is laid. The key to good brick laying is in the generous use of water. Bricks are highly absorbent, and, if laid up dry, will absorb water from the mortar before the bond has been completed. Besides a thorough soaking of the brick a few hours before its use, the mortar should be mixed with the maximum amount of water that it is possible to use and still produce a workable mix. The compressive strength of mortar increases with an increased proportion of cement and/or with a decreased water-cement ratio. The tensile bond strength between mortar and brick, however, increases with the increased flow of the mortar, that is, with an addition of maximum water. However, in most masonry structures the compressive strength of the mortar is of secondary importance to the bond strength, to water retentivity, to workability, and to low volume changes, like shrinkage. In earthquake and high-wind areas where resistance to lateral forces is important, the strength of a wall depends primarily upon the strength of the bed joints through which failure normally occurs. A good mortar proportion for brick work consists of 2 parts cement, 1 part fireclay or hydrated lime, and 9 parts graded sand.

Concrete masonry blocks should be laid with a somewhat drier mortar, and, unlike clay brick, the blocks themselves must be kept dry before and during their placement in a wall. For ordinary service, a mortar proportion of 2 parts cement, 1 part fireclay, and 6 parts sand has proved to be generally satisfactory.

To achieve optimum mortar strength and workability, an owner-builder should experiment with various mortar proportions. A high cement content in the mortar produces considerable strength, more rapid set, only fair workability, some shrinkage in drying, and little elasticity. A high lime content, on the other hand, produces less strength, slower setting, easy workability, considerable shrinkage, but good elasticity. A high sand content produces little shrinkage, less strength, and poor workability. A high water content produces easy application, less strength, and excessive shrinkage.

The first course of concrete masonry can be set either directly onto a freshly poured foundation footing or onto a bed of mortar above it. In either event, one should first accurately set the corner blocks and adjust all intermediate units to fit both the wall length and the necessary mortar spacing. This last process is known as running out the bond. Little difficulty will be encountered if the block sizes conform to the 4-inch module, and if the wall lengths are designed with this modular coordination in mind. On this first course, the blocks should be carefully checked for correct alignment and for plumb. The use of a heavy, nylon line is essential for keeping accurate alignment and level.

After the first course is laid around the perimeter of one's construction, every inside or outside corner is laid up, usually four or five courses higher than the center of the wall. As the corner course is laid, it must be checked with a level for alignment, for being plumb, and for being level. Furthermore, each block should be carefully checked with a level or a straight edge to insure that the faces of the blocks are in the same plane. The use of a story pole, which is simply a board with markings for the height of each course, provides accuracy in establishing the top of each masonry course.

The next step to building a masonry wall is to lay blocks between the corners. The top and outside edges of these stretcher blocks are kept in line with a taut plumb string. A little practice in applying mortar and setting blocks will make it possible to make adjustments

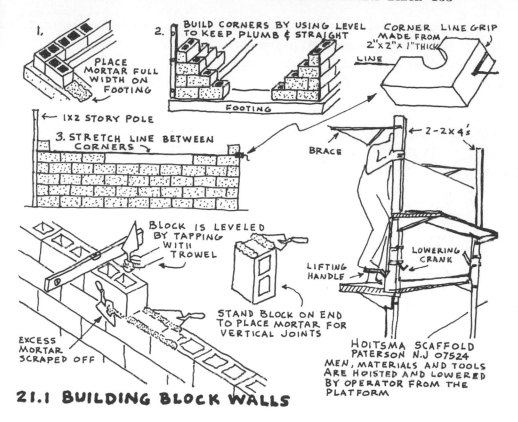

1. PLACE MORTAR FULL WIDTH ON FOOTING

2. BUILD CORNERS BY USING LEVEL TO KEEP PLUMB & STRAIGHT

FOOTING

CORNER LINE GRIP MADE FROM 2"x 2"x 1"THICK

LINE

← 1x2 STORY POLE

3. STRETCH LINE BETWEEN CORNERS

BRACE

← 2-2x4's

BLOCK IS LEVELED BY TAPPING WITH TROWEL

STAND BLOCK ON END TO PLACE MORTAR FOR VERTICAL JOINTS

LIFTING HANDLE →

LOWERING CRANK

EXCESS MORTAR SCRAPED OFF

HOITSMA SCAFFOLD PATERSON N.J 07524 MEN, MATERIALS AND TOOLS ARE HOISTED AND LOWERED BY OPERATOR FROM THE PLATFORM

21.1 BUILDING BLOCK WALLS

of final position while the mortar is soft and plastic. Attempts at adjustment after the mortar has stiffened will most certainly break the mortar bond and allow the penetration of water.

Watertight mortar joints, as well as a neat, finished appearance, depend upon proper tooling when the mortar is set. This tooling operation compacts the mortar and forces it tightly against the blocks on both sides of the joint. This is best done with a concave or V-shaped tool as soon as the mortar has become thumb-print hard.

A wide variety of blocks that conform to the 4-inch module are now available and are designed to contend with just about any conceivable wall-construction problem. Of special value is the header block, which avoids the former necessity of building wooden bond-beam forms. A block that receives the side flanges of metal window and door frames is also available. Even window-sill blocks are made, eliminating the need for any wood trim in a fireproof, all-masonry building.

A number of masonry-block manufacturers now sell the mortarless block. In my judgment, this block is much better suited to use by the average owner-builder. It can be laid faster, with less need for skilled labor. As a general rule, the mortarless block is more precision-engineered than the common variety of block, which must be laid up with mortar. A variety of joining designs have been developed, including tongue-and-groove, ship-lap, ball-and-socket, and so forth. In practice, one lays the first course in the wet foundation footing around the perimeter of the house. Obviously, it is important to get this first layer level, in line, and plumb, since the erectness of the rest of the wall is determined by the correct alignment of this first course. Blocks are then set layer upon layer until a desired height is reached. No mortar is used and the blocks fit in one position only. When the final height is reached, cores at the corners and cores on 2-foot centers are poured with concrete.

It has been estimated that the cost of masonry construction accounts

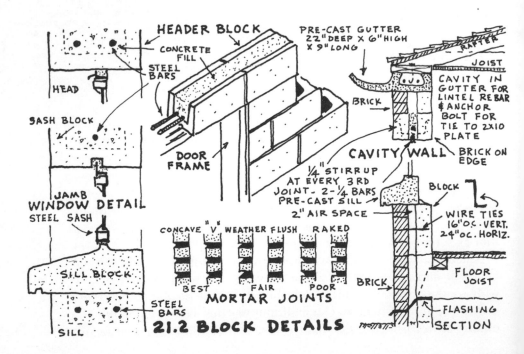

21.2 BLOCK DETAILS

for more than 5 cents of every construction dollar spent. This fact may have spurred the National Concrete Masonry Association to make an in-depth investigation of the restraints on masons' productivity. The study was published in 1973, and final recommendations by the Association to the masonry industry prove to be interesting reading.

First, it was found that the larger the unit block size, the greater one's productivity. Laying a 6-×-8-×-16-inch concrete block is 3.29 times more productive than laying a standard brick. Too, a 20 percent increase in efficiency is achieved by using a tower-scaffolding system, instead of using the conventional tubular-frame scaffolding. The tower scaffolding, illustrated in Figure 21.1, has a continuous, incremental-adjustment feature which keeps the mason positioned at the ideal work level, and which minimizes his fatigue by reducing the bending required of him. Another 10 percent increase in mason productivity resulted from the use of corner poles. Corner poles, illustrated in Figure 21.4, are plumbed and trued outside of the block wall, and they double as story poles to maintain the level of each course.

Complete corner-pole fittings and telescoping braces can be purchased mail order from Masonry Specialty Company, 172 Westbrook Road, New Kensington, Pennsylvania 15068. Corner-pole fittings cost $13 a pair and the telescoping braces cost an additional $15 a pair. Every tool possibly required by an amateur or a professional mason can be ordered through this company's illustrated tools and equipment catalogue.

Perhaps the greatest achievement for increased productivity, some 70 percent greater efficiency, came as a result of Association experiments in surface-bonding masonry walls. Surface bonding is a relatively recent development. It was developed and first used by the U.S. Army Corps of Engineers in 1967. Since then, a number of concrete-products companies have capitalized on the original research by the Corps. Bonsal Company, for one, has a U.S. and foreign patent pending on Surewall surface-bonding cement. It would certainly be a waste of money for an owner-builder to purchase fancy-packaged bags of this product when the same material can be formulated directly from the Corps of Engineers' research and from the raw components.

The cement-bond material is applied to both sides of concrete block walls, which have been stacked without mortar. The bonding mixture creates a bond having greater racking strength than conventionally mortared walls. Racking strength is a measure of the ability of walls to withstand sheer forces. Not having to use mortar also eliminates most of the time-consuming, skilled labor required in block laying.

As well as providing stronger wall joints, surface bonding tends to adequately waterproof walls. One of the most attractive features of the bonding mix is its low cost. On the basis of a recommended $\frac{1}{8}$-inch-thick application, the material costs 5 cents a pound or 3 cents per square foot of wall area. Weights of the ingredients needed to make a 25-pound batch of the bonding mix are as follows:

Cement	$19\frac{1}{2}$ pounds
Lime	$3\frac{3}{4}$ pounds
Calcium stearate	$\frac{1}{4}$ pound
Calcium chloride	$\frac{1}{2}$ pound
Glass fiber	1 pound
Water	$1\frac{1}{2}$ gallons

Mix the dry ingredients thoroughly in one batch, and the calcium chloride and water in another. The liquid is added slowly to the dry ingredients and, again, is thoroughly mixed. To get maximum bondage, the mix should be as thin as possible to be handled with a trowel. The block wall should be clean and thoroughly dampened before applying the surface bond. Much troweling, however, weakens the bond. It is important to water cure the coated wall to prevent rapid dehydration and cracking.

Another major improvement in the masonry industry occurred when they started making lightweight blocks, using cinder, pumice, basalt, perlite, and so on. Insulating and sound-absorbing qualities are higher in these blocks, and, with less weight involved, the units can be laid faster.

One would naturally suppose that the value of a building material depends less on its appearance than on its structural and insulative value. But modern taste has dictated unreasonably high standards for exterior appearance. In some circles, the degree of affectation supersedes the more basic considerations of cost and strength. In the 1930s, attractive facebrick was first extensively used as a front-wall cover up, and, for a while, about every other new house suffered from this "veneereal" disease. In more recent years, the trend has been toward masonry grilles and patterned-block units. These units are manufactured in virtually any color, shape, size, or texture. One can buy split blocks, slump blocks, or scored blocks, depending upon the effect one wishes to achieve. There is also an equally wide range of patterns, depending upon how one treats the mortar joint or how one combines the blocks in any one of a hundred different ways.

None of these decorative features compensates for the fact that concrete blocks provide little resistance to heat loss. An insulated block wall built of lightweight aggregate offers a resistance factor of 2.94. Condensation will occur on the inside surface of this wall when the inside temperature is 70°F, relative humidity is 70 percent, and the outside temperature is 20°F. High humidity is not uncommon in bathrooms, kitchens, and utility rooms. The resistance value of a lightweight block wall can nearly be doubled to 5.79 by filling block cores with expanded mica. A very high resistance rating, 9.13, is achieved by bonding 1-inch polyurethane-foam insulation board onto the inside surface.

In contrast to the growing number of builders who go to absurd lengths to find new ways to use concrete masonry for surface effect, an equal effort is being made by research institutions to develop methods whereby house-needy families may build their own homes out of masonry units produced by themselves—masonry units for structures and enclosures rather than for looks. The Tuskegee Institute engineering staff was fifteen years developing a concrete block wall-and-building system. Said F. D. Patterson, Institute president:

> Our experiments in housing have been oriented toward farmers because they have the greatest need in this area for housing improvement, but are least able to provide themselves with the improvement needed through normal commercial arrangements.
>
> Experimentation at Tuskegee Institute has revealed that building costs can be substantially reduced by utilization of labor of the farm family in seasons when there is little demand for labor in farm production. Without specific training and under minimum supervision, unskilled labor in home construction can be performed by members of the family. Short courses have, in a brief period, prepared farmers for satisfactory performance on semiskilled jobs in home building. In the Tuskegee Institute experiments, reduction

in material cost and in labor cost have made possible convenient and attractive homes for farmers, with the expenditure of surprisingly small amounts of cash.

Architect Frank Lloyd Wright was one of the first to develop an improved concrete-block system. His Usonian Automatic block has the singular quality of providing a continuous air space between the inner and the outer wall. This air space is, of course, invaluable for insulation against heat, cold, and moisture. The Alcon block is another type of twin-wall block which successfully incorporates a motionless or dead-air space between the outer and inner walls. No direct, horizontal contact is made between the blocks. This 16-inch-square unit can be mass produced on the site by the owner-builder, using a simply constructed metal gang form.

Absence of a continuous air space in conventionally built 8-inch brick walls is, perhaps, the main reason, insulation-wise, why this type of wall is becoming less prevalent. Experience has shown that, in instances where the brick header is continuous throughout the full thickness of the wall, extreme care must be exercised in construction in order to obtain a wall that will resist the penetration of heavy rains accompanied by high winds. Concrete block or structural clay tile is usually used to back up the clay-brick facing. As is often the case where

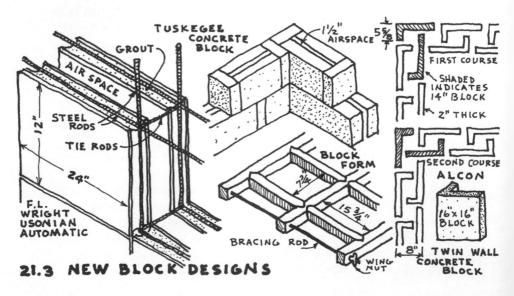

21.3 NEW BLOCK DESIGNS

different materials are used (concrete and burnt clay in this above-mentioned instance), the thermal- and moisture-expansion coefficients of facing and back-up are different. Movement of the concrete-block back-up, due to shrinkage as the wall dries out, creates an eccentric load on the brick headers, which tends to rupture the bond between the headers and the mortar on the external wall face.

The logical solution to this problem of moisture penetration and poor insulation is to separate the inner wall from the outer one. The cavity wall has been used in England for fifty years or more, but has only recently gained favor in this country. New York's nineteen-story Veterans' Hospital is claimed to be the highest cavity-wall structure in existence. The wall consists of an outer layer of facing brick and an inner layer of concrete block with 2-inch, continuous air space between the two. The two surfaces are secured by Z ties, spaced 16 inches apart vertically and 24 inches apart horizontally. The cavity is drained of any penetrating moisture by weep holes to the outside.

Where a resistance to heat flow is required, in addition to the insulation provided by an air space, insulating material can be installed directly into the cavity. The wall then has the double advantage of the lower heat gain typical of clay products and a high resistance to heat flow due to the cavity stuffing. According to tests at the Armour Research Foundation and Pennsylvania State College, the pouring-type insulation (S.C.R.) produced by Owens-Corning Fiberglas proved superior to all other varieties.

Cheaply built brick buildings suffered perhaps the greatest structural devastation in the 1933 Long Beach, California, earthquake. The human misery and financial loss occuring with this event prodded the masonry industry to find safer and stronger ways of building with brick. So now we have two new methods of brick construction: reinforced brick masonry and reinforced grouted brick masonry. For a wall height of 20 feet, an unreinforced thickness would require the wall to be 12 inches thick, whereas a reinforced thickness would require a wall of only 8 inches. That is, the unreinforced wall requires 50 percent more brick, and 50 percent more weight must, therefore, be carried by the supporting structure. A brick or block wall reinforced throughout every foot in height with Dur-O-Wall steel-masonry wall reinforcement (Box 628, Syracuse, New York) has a 71 percent greater flexural strength than its unreinforced counterpart.

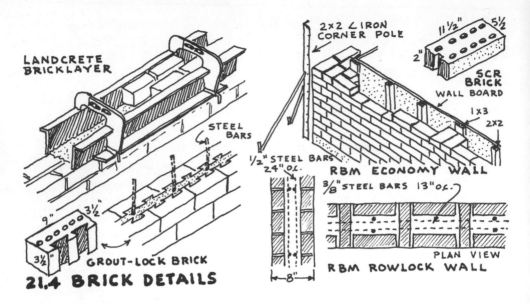

21.4 BRICK DETAILS

Along with the many varieties of fabricated metal reinforcements, specially designed brick, like the Grout Lock and S.C.R., are now manufactured for those who want to insure maximum bond with the mortar bed. Actually, however, the same structural results can be achieved using common brick and standard-size reinforcing bars. This method is referred to as Rowlock construction. This system of reinforced brick masonry construction offers substantial savings in speed of erection and in the amount of material used. Perhaps the most dramatic use of this system of construction is in the Lunt Lake apartment building in Chicago. Walls of the first story are only 10 inches thick, and the remaining eight stories are 8 inches thick. This remarkable wall strength is achieved despite the fact that all the bricks in this building are laid on edge, a pattern typical of Rowlock bond. One of the developers of Rowlock, Henry Holsman, writes in the June 1949 issue of *Brick and Clay Record:*

> If brick masons and manufacturers will adopt and promote this type of reinforced brick construction, it will put the mason and the brick industry on an equally competitive footing with other methods of building construction, if not far in advance of them. In my judgment it can be developed to make the concrete skeleton

method of construction quite obsolete. To abolish the temporary forming in building construction, which sometimes and quite often costs as much as or more than the structure itself, will be a great economical boon to the general public who, after all, must pay the extra cost of any inefficient or unnecessary method of building construction.

Perhaps the lowest-cost brick-wall construction method, appropriately termed "economy wall," was developed for the Lee County Housing Authority, Tupelo, Mississippi, during the Second World War. The wall construction consists of nothing more than a 4-inch brick wall built on a reinforced brick grade beam, which in turn is supported on 8-×-16-inch brick piers. The contractor's building cost of $1,538 for each two-bedroom unit was lower than competing bids using frame construction. This type of construction well illustrates the fact that, from a design standpoint, the outside layer of a brick-masonry wall should not function as a mere skin or covering. It should serve as part of the structure as well. This is a fundamental consideration which should be pursued with the choice of any building material. Finish should be an active, load-carrying participant in the construction rather than a passive, useless veneer.

The final stage of any masonry construction involves the cleanup of random smears of mortar or traces of the white efflorescence that often appears on fresh brickwork. It is best to wait a few weeks after the last brick is laid before applying a 10 percent cleaning solution of muriatic acid. A stiff brush or rag is best for applying the acid to the wall. A thorough wash down with water should precede and follow the acid treatment. This will be about the last thing that need ever be done to the owner-built masonry home, at least during the builder's lifetime. There is a real comfort in knowing that one's masonry home has a pleasing appearance and structural durability. Good insulation and fireproofing, from a material that is relatively cheap and convenient to procure, are features that make a masonry home a definite candidate for selection by the amateur owner-builder.

22 CONCRETE

Concrete continues to be the most basic of all of the poor man's building materials. While other processed building materials have increased in price many times over in past decades, the price of cement and aggregates has remained relatively stable. For a ten dollar bill one can produce a cubic yard of material, which will virtually last forever without maintenance. Concrete is a fireproof material which can be molded into any desired form or to any surface texture. It is an inexpensive material which offers great compressive (loading) strength. With the addition of a few dollars for steel reinforcement, considerable tensile strength is added, reducing concrete thickness and dead weight. With minimum cash, but also with an understanding of the nature of concrete and with a few tools and techniques for working it, the owner-builder can expect only the best results from this marvelous material.

If by "poured concrete" one means the hard work of building wooden forms, mixing, and shoveling (or pouring) concrete into position and eventually removing the wasted lumber of the forms, then we are, indeed, talking about two different things. With barely an exception, conventional uses of concrete are grossly inefficient, wastefully expensive, and laborious. Yet, even professional cement workers have become so accustomed to this established procedure that they fail to consider the possibility of a better way to accomplish their concrete tasks or to see the wider potentialities for this material. The house-building industry includes as many archaic, unscientific, and unimaginative methods of construction as do any other industries in their respective fields. The knowledge of materials, the design and structure that does filter out of alert research channels, completely fails to reach the small-homes builder or the individual owner-builder.

By way of contrast to this stultifying situation, the work of Robert Maillart immediately comes to mind. Maillart was primarily a bridge designer in Switzerland. No two Maillart bridges have the same design or are built with similar structural solutions. His was an inquiring mind, continually searching both for pleasing line and for economy.

A beautiful illustration of Maillart's economy-of-means concept is seen in the manner in which he eliminates the superfluous by integrating all structural components. Every cubic foot of concrete is used structurally in the satisfaction of tensile and compressive stresses. In one railroad bridge, the rails, themselves partially embedded in the concrete, function structurally as reinforcement for the bridge. Even the handrail on one pedestrian bridge is designed to function structurally.

Maillart did not confine himself to the design and construction of bridges alone, however. His pavilion at the 1939 Zurich Exposition is a historic architectural masterpiece. Although over 50 feet high, the arched concrete shell of the pavilion is nowhere over $2\frac{1}{4}$ inches thick.

As might be expected, the impact of Maillart's work is being felt only slowly, yet he was one of the earliest builders to employ a reinforced concrete shell as an enclosure for space. This was a direct challenge to the mathematical theory of elasticity, which exhorts one to employ mere mathematical procedures for structural analysis. Except for the simpler type of dome, doubly curved dome surfaces are practically impossible to represent on paper. The theory of elasticity has long inhibited the normal development of structural knowledge and experience in the field of concrete construction.

Professor Pier Nervi of Italy and Felix Candela of Mexico are two architects who did carry on Maillart's work on rigid-shell construction.

In some of Candela's hyperbolic paraboloids, the stresses are so low that steel reinforcement becomes superfluous. Concrete thickness may even be reduced to as little as $\frac{3}{4}$ inch. Nervi developed a ferro-cement process, which made it possible to reduce concrete thickness to $\frac{1}{2}$ inch. Several successful efforts were made by Nervi in the early thirties to build ferro-cement boats. Both the hull and the cabin of these small boats were less than 2 inches thick. These amazingly strong skins were achieved by forcing a high-quality cement mortar through several layers of steel mesh.

More recently, 40-foot-long ferro-cement canoes have been built in Canada, weighing slightly over 100 pounds. The hulls of these boats are about $\frac{3}{8}$ inch thick, combining fiberglass and woven-steel-wire fabric, layered with a mix of 10 parts Portland cement, 3 parts acrylic latex, and 1 part water.

From houses to bridges to boats—in each case the search is for simple structure through an economy of means. Reinforced concrete makes such a visionary search quite likely to bring success in building. This new direction in structural design must grow from experimentation or from empirical knowledge. Mathematical calculation, especially when some experiential factors are not taken into consideration or are unknown, can only restrict and limit an imaginative solution. Simpler solutions are adequate when the designer is a builder and not merely a computer.

Of all the many characteristics of concrete, the one that can be most useful to us in contemplating building design and a system of structure is its property of being monolithic. Concrete itself has high strength under compression but low strength in tension. Actually, the tensile strength of concrete is about one-tenth its compressive strength. Where the compressive strength is greater, the tensile strength is on the order of one-twentieth of the compressive strength.

The fact that the coefficient of thermal expansion for concrete and steel is practically the same makes it possible to use steel to achieve high tensile strength and ductility when it is combined with the weaker concrete. The affinity for bond between concrete and steel ensures an effective cooperation between them when under load.

Stone masonry, as was discussed in chapter twenty, has a compressive strength that is many times greater than its resistance to tension. It is poorly adapted to bending stresses, as when used over an opening.

Wood, on the other hand, offers about equal resistance to compression and to tension in the direction of its fibers. It is, for this reason, more widely used in building construction.

Wood and stone construction alike lack that important property of being monolithic, a property peculiar to reinforced concrete. For instance, a wood floor or roof consists of joists and boards with no structural cooperation between these elements—only a dead loading of the boards on the joists. With a concrete roof, the slab transfers loads at right angles to the span of the beams. It also acts like the flange of a T beam, transmitting loads in the direction of the beam span as well. This is an application of the principle of continuity.

Before getting too involved with applications, it may be wise to remind ourselves that concrete, used as a building material, has certain practical limitations as well as virtues. Strength is not necessarily the main factor to be sought in building monolithic structures. Concrete with high compressive strength has susceptibility to shrinkage cracking. The problem of shrinkage is perhaps the major hurdle to overcome when building with concrete. One way to do this is to use a leaner mixture of cement, with accruing lower cost. In terms of amount used, only enough cement to protect the reinforcement against corrosion is actually necessary. Man has associated strength with mass long enough. We need to develop some spatial concepts which leave room for stable reactions from thin-shelled structures.

Without doubt, the water content in concrete mixtures is the most important ingredient. Fifty thousand tests made at the Jarvis Institute in Chicago showed that very small variations in water content produce more important variations in concrete strength than similar changes in other ingredients. It was demonstrated that, in certain instances, a 1-part-cement to 9-parts-aggregate mixture is as strong as a 1-part-cement to 2-parts-aggregate mixture, depending only upon the water content of each mix. The absolute minimum amount of water should be used in concrete mixing to obtain optimum strength. It is far better to tamp and to vibrate the concrete than it is to allow it to flow freely.

Concrete strength depends upon a few other factors besides mixture proportions. A higher compressive strength can be achieved when the concrete is allowed to mix several minutes after it is uniform in appearance, with all of the ingredients uniformly distributed. Unknown to most builders, the ultimate strength of concrete depends in some degree upon the mixing sequence. The Rocla Laboratories of Melbourne, Australia, found that an 11 percent higher tensile strength can be obtained when sand and cement are first mixed together, with the coarse aggregate being added after the water. This is an 11 percent gain in strength over the customary practice of mixing coarse aggregate in with the dry sand and the cement.

When all these slight increases in strength are made through proper mixing practices, the use of minimum water content, and thorough settlement and compaction, the proportion of cement can be reduced, thereby lessening tendencies to shrinkage cracking. These important features will be appreciated once the actual forming technique is employed.

Several times during the past fifty years, truly major advances were accomplished in developing techniques for pouring low-cost concrete structures. These developments were made by builders who, through long experimentation and experience, came to understand the actual nature of concrete construction.

The world-famous inventor Thomas Edison was one of the earliest to investigate the possibilities poured concrete offered to the low-cost builder. In 1907, Edison prophesied this type of housing through widely publicized channels. Later, in 1915, he patented a system for pouring a complete residence in one massive form.

Then, in 1917, Van Guilder improved on Edison's cumbersome technique. He patented a double-wall form, which allowed three to four courses of wall, each 10 inches high to be cast a day. His form consisted of nothing more than four flat steel plates for the form walls, held in position by a toggle joint and operated by a hand lever. When the toggle was released by the lever, the outer plates moved out and the inner plates moved in, permitting the form to be lifted clear of the poured walls. For corners a special mold was used. Thousands of Van Guilder homes were built immediately following World War I.

For a brief period after World War II, interest revived in the reinforced-concrete, hollow double wall. A number of companies began to produce machine forms à la Van Guilder. Builder John Geiger of Churubusco, Indiana, was about the only person to make major improvements in the Van Guilder method. The principle of Geiger's patented form is somewhat different from the pour-and-release systems previously used. His lightweight aluminum form was designed to slide

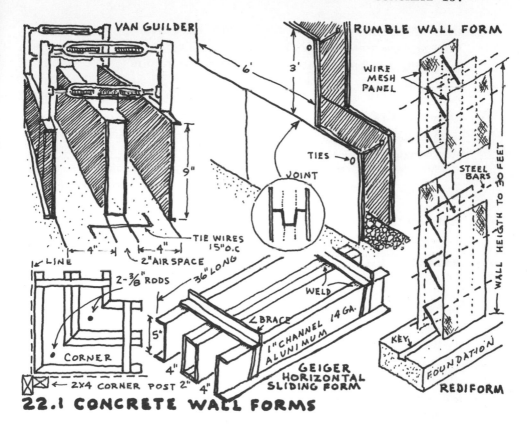

22.1 CONCRETE WALL FORMS

horizontally, forming 5-inch-high double wall courses, a layer at a time. For top efficiency, three people are needed to operate this building system, with one person sliding the forms, one person shoveling concrete into the forms, and one person mixing the concrete. As much as 200 square feet of wall per man per day can be built with the Geiger forms.

It is very important to maintain a consistent concrete proportion. A 1-part-cement, 3-parts-sand, and 3-parts-gravel mixture (under $3/4$ inch in size), with less than 5 gallons of water for each sack of cement, proves to be the ideal proportion. If the concrete is too wet, it will slump when the form is pulled ahead. If it is too dry, it will require light tamping. There should be a slight slumping of the concrete in the wheelbarrow as it is wheeled from the mixer to the construction site. With the correct water content, no form marks will be left in the wall when the form is pulled ahead. Remember, in order to mix sand and cement to a dry consistency, add gravel after water, and be sure to compact the mix thoroughly before using it.

Skeptics who would feel insecure living in a house having 4-inch concrete walls should read some of the findings of tests made at the University of Illinois. A 4-inch-thick concrete wall was found to have, under axial loading, an ultimate compressive strength of 80,000 pounds per lineal foot. This proves to be a very high factor of safety. The flexural strength—that is, the resistance of the wall to wind or to other transverse loading—of a hollow double wall with a wind pressure of 25 pounds per foot was found to have a safety factor well over the factor of 2.

According to the American Society of Heating and Ventilating Engineers, the heat transmission coefficient of a hollow, 4-inch concrete wall having an 2-inch air space has been determined to be 0.43 Btu. When the cavity is filled with rock wool, the heat loss coefficient is reduced to 0.15 Btu. This is to be compared to a 0.30 Btu coefficient for standard wood-frame walls.

When a solid concrete wall is desired, one can choose a number of different patented forming devices, all of which reduce conventional, wooden forming procedures to zero. The single-thickness, monolithic concrete wall is unsatisfactory for housing, since the high capillary action of concrete allows moisture penetration and since the insulative value of such walls is low. In order to overcome the obvious shortcomings of concrete insulation and, yet, to take advantage of the less expensive methods of forming single-thickness walls, a totally new concept of concrete mixing has been developed. The mixture is termed "no-fines," simply because no sand aggregate is used in the mix. O. G. Patch, a concrete technologist for the Bureau of Reclamation, is given unofficial credit for being the creator of the no-fines mixture. Experiments in its use were conducted at the Grand Coulee Dam in the 1930s.

The object of this mix is to obtain air space for the aggregate. Proper water content is crucial to the successful no-fines mix. Water should be sufficient to assure a thorough coating of the gravel with cement paste, but not enough to make the mixture flow and to blind the pores. The normal water requirement for a 1:8 mix will range from $3\frac{1}{2}$ to 4 gallons per bag of cement. A proportion of cement, water, and uniformly sized aggregate, which passes through a $\frac{3}{4}$-inch sieve, are the only raw materials required. The cement saving is greater with no-fines, for as little as 1 part cement is used to 10 parts aggregate. This, of course, reduces both the tensile and the compressive strengths of the wall, yet a 6-inch-thick no-fines wall is still adequate for a two-story residence, providing the walls are not subjected to unusual, eccentric loads or to excessive binding stresses, such as might occur with slender piers or with the bearing of heavy lintel.

The drying shrinkage of no-fines concrete has been found to be roughly one-half that of regular concrete. More important yet is the fact that no-fines is two-thirds lighter than regular concrete. This means that, during pouring, the hydrostatic pressure on formwork is considerably lower, about one-third lower—enabling use of a relatively light, easily assembled form.

In 1950, the Southern Rhodesia National Building and Housing Board built some 1,000 no-fines concrete houses. Floors in these units, poured 3 inches thick, were a no-fines mixture. Six-inch no-fines walls were covered, inside and out, with thin plaster and whitewash.

A 10-inch no-fines wall, with its surface protected in similar manner, has a heat transmission value of 0.30, equivalent to an 11-inch-cavity brick wall. The open-textured mass of no-fines concrete provides little bonding strength, and, for this reason, walls are usually designed to avoid using steel reinforcement. Consequently, heavy, eccentric loadings and long spans must be eschewed.

The low hydrostatic pressure of no-fines concrete, mentioned above, permits the use of large, lightweight forms. Forms may also consist only of open-steel mesh or of expanded metal on a light, timber frame work. RediForm (Box 736, Austin, Texas 78613) is a relatively new, commercial forming system, ideally suited to a no-fines installation. It consists of two "skins" of metal lath, separated by a steel truss. When poured with concrete, the forms become an integral part of the finished product, as illustrated in Figure 22.1.

Portable, monolithic wall forms of the type invented by the Magdiel brothers are suited to a no-fines as well as to a regular concrete wall. The Magdiel form, already mentioned in the chapter on earth-wall construction, has a number of ingenious release devices, which make it an efficient machine for forming concrete foundations and walls. Another steel wall form of merit has been developed by a South

African builder, Roy Rumble. The Rumble form is 3 feet wide and 6 feet long. It is constructed of light-gauge steel held in place by ties inside the wall. The system of forming is quite unique and simple. After the first section is poured, a second set of forms is placed on top of the first form, and it is joined in a tongue-and-groove assembly. The upper section is then poured, after which the lower forms are removed and set on top of the upper set of forms for the next pour. A "pebbled texture" effect can be created on the outer wall by placing wire mesh against the form.

23 PRECAST CONCRETE PANELS

Special commendation should be made for the work of numbers of well-qualified architects, engineers, and builders who have burdened themselves with extra effort to solve some aspect of the low-cost housing problem—in preference to concentrating their efforts on lucrative private practices. It was not monetary gain but the challenge to devise a means for constructing an inexpensive, fireproof, all-concrete house that motivated Thomas Edison in this work, for he lost money with his experimentation.

Edison's diligent activity set the pace for the work which was required to establish the concept for the precasting of concrete wall panels, which is the most viable solution to present-day needs for mass housing. Structural engineer F. Thomas Collins devised an ingenious system of fabricating precast units. Milwaukee industrialist Paul Juhnke has spent many years developing an owner-built, tilt-up concrete panel. Architect Alvaro Ortega, one-time staff member of the United Nations Technical Assistance Board, designed a low-cost housing project of 200 homes in Bogota, Colombia. These $650 units, built in 1955, utilized a unique so-called "vacuum-process" method of precasting concrete roof and wall units.

Precasting eliminates the wasteful use of form lumber. Broadly speaking, about one-third to one-half of the total cost of a reinforced concrete structure is in the wooden-form work itself. Precasting also saves on the vertical and horizontal movement of concrete batches to various pour locations on the building site. Through the use of his vacuum technique, Ortega establishes another advantage, a saving on form work altogether. When the first slab is poured, vacuum pads are applied which draw out excess moisture from the concrete. This makes the slab prematurely hard, and its surface dries rapidly. Then a sheet of building paper is spread over the slab, and a second wall or roof section is cast on top of the first slab. In all, eighteen slabs are poured, one on top of the other.

Alvaro Ortega is currently associated with the Minimum-Cost Housing Group at McGill University, Toronto, Canada. One of his present projects involves experimentation with building blocks made from waste sulphur.

Tilt-up methods have been used successfully since 1946. Tilt-up is a form of on-site, precast construction in which the walls are cast, usually on a floor slab, in a horizontal position. They are then mechanically tilted up to a vertical position and are set in place to become an integral part of the completed structure. Corners and intermediate joints are poured in place. Column-type footings, located where precast wall panels join one another, can be used in place of the expensive, continuous, reinforced foundations required of conventional, masonry buildings.

With this method of construction, the problem of placing the slabs in position may appear to be a major drawback for the individual owner-builder, especially when the units are large wall or roof sections. However, as illustrated below, a simple and inexpensive, hand-operated tilting mast can be put together by the owner-builder. An even more efficient tilting boom can be operated from the controls of a small, wheel tractor.

One such tractor-powered, tilt-up, concrete structure was built a few years ago at Texas A & M University. The wall panels, a maximum size of 10 feet square, were poured 4 inches thick with a mixture of 1 part cement, $\frac{2}{3}$ part water, $2\frac{1}{4}$ parts sand, and 3 parts coarse aggregate. After a three-day curing period, the panels were tilted to a vertical position with a tilting frame made of 2-inch pipe. Once in

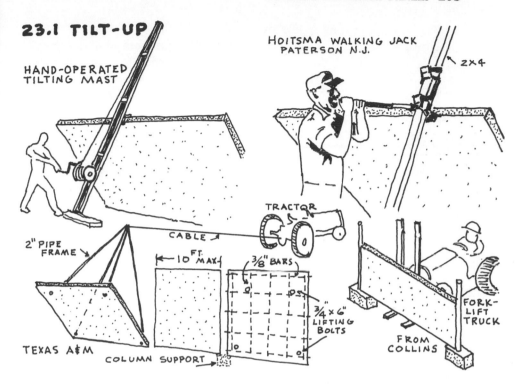

23.1 TILT-UP

HAND-OPERATED
TILTING MAST

HOITSMA WALKING JACK
PATERSON N.J.

2×4

TRACTOR

2" PIPE
FRAME

CABLE

10 FT. MAX.

3/8" BARS

3/4" × 6"
LIFTING
BOLTS

FORK-
LIFT
TRUCK

TEXAS A&M

COLUMN SUPPORT

FROM
COLLINS

place, the panels were braced, and an H-shaped, cast-in-place column connection was poured between adjacent panels.

A number of factors contributed to the extremely low building cost of $1 per square foot of this experimental, 20-×-40-foot implement shed built at Texas A & M University. For one thing, it was determined that only a 10-inch-square footing needs to be built under each panel end. It was also found that the simplicity of the construction of these panels made it possible to employ unskilled labor. Wall panels are made up in a horizontal position in standardized forms.

Tilting or lifting a precast concrete panel into its vertical position imposes maximum stresses on the panel itself, early in its life and before the concrete has attained high tensile strength. On the Texas A & M project, it was found that judicious selection of the location for the pick-up points on the finished panels helped to minimize the induced flexural reaction.

A smaller size panel, of course, requires a mechanism for lifting and for transport that is less extensive. Collins used a fork-lift truck to set his 8-foot-long panels into position. Juhnke rented a tow truck to transport and to place his 44-inch-wide panel units.

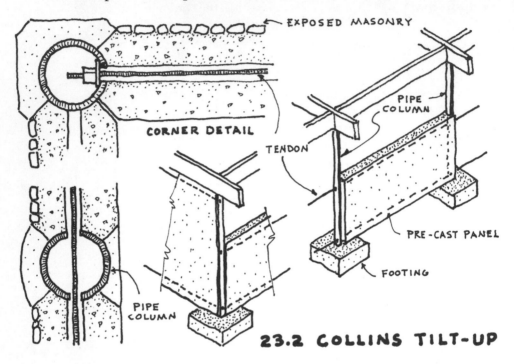

EXPOSED MASONRY

CORNER DETAIL

PIPE COLUMN

TENDON

PRE-CAST PANEL

FOOTING

PIPE COLUMN

23.2 COLLINS TILT-UP

Both Collins and Juhnke precast their wall panels off site. One advantage of doing this, as Juhnke points out, is that the panels can be made under cover during spare time and in inclement weather before the project begins. Each of the lift-slab constructions, illustrated in Figures 23.2 and 23.3, have their own distinct advantages. Collins first set concrete piers every 8 feet around the perimeter of his building project. In the center of each pier, he accurately located a vertical-pipe column. He then proceeded to fork lift the panels to positions between pipes. A length of pipe is used for the side of the form when the panels are poured, so that, when the finished panel is set between the vertical pipes, which are already erect and in place, a snug fit is made between the precast panel and its supportive pipe member. Smaller, horizontal pipes are also set in the castings, one near the base of the panel and another near the top. When the panel is finally in position, steel tendon rods are fed through these smaller, horizontal pipes in the panels and through holes in the vertical-pipe columns. At opposite corners of the building, the tendon rods have threaded nuts which, when tightened, literally squeeze the wall panels together— one at floor level and another near the ceiling. Collins describes this corner anchoring and the drawing together of the wall panels as a post-tensioning of the tendons. To gain this structural effect, it is necessary to pressure grout the pipe columns and the preformed holes after the tendons are tightened. The floor slab is poured after the wall panels are in place and post-tensioned.

Juhnke prefers to work with a slightly smaller size panel, only 44 inches wide and 8 feet long. His joining system is also quite different from that mentioned above. The vertical space between panels, which is 4 inches wide and 5 inches (the depth of the wall) deep, is cast with concrete after the panels are set in place. The problem of how to properly join panels at wall corners has not been adequately explored by Juhnke. Ideally, there would be no right-angle corner problem if the panels could be placed in a continuous curve, as shown in the expandable-house design, Figure 10.3.

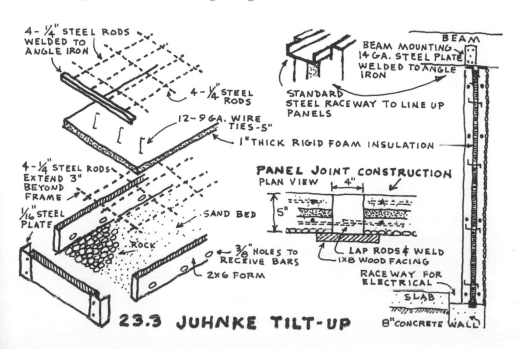

23.3 JUHNKE TILT-UP

Owner-built, tilt-up walls are best poured on a flat sand bed in a 2-×-6-inch framework. An even layer of 1-inch-diameter stone is laid on the sand and is covered with 2 inches of concrete. Then, a gridwork of $\frac{1}{4}$-inch steel bars is embedded into the wet concrete. A 1-inch layer of rigid foam, polystyrene, is placed directly on top of the concrete and is covered by an additional 2-inch pour of concrete. Wire ties, extending through the foam insulation, sandwich together the two layers of concrete after its initial set. The final, 2-inch layer of concrete should also have a gridwork of reinforcing bars. Material cost for a 44-inch-by-8-foot panel is about $10, with concrete and steel costing about $6 and foam costing about $4.

For this nominal investment and a small amount of work, one realizes in return a permanent, maintenance-free, well-insulated, and fireproof wall section. The only thing left to do is to erect the panels, weld the ends of the steel bars together, and cast the connection joints.

Single-span, rigid-frame members can be precast by owner-builders with little trouble. Forms are either self-fabricated or are purchased from companies which specialize in building forms for precast-concrete products, such as R. L. Spillman Co., Box 7847, Columbus, Ohio 43207. Investment in commercial forms would only be feasible where extensive group building programs are anticipated.

The concrete panel-building techniques mentioned in this chapter are seldom employed by orthodox builders or by homeowners. The latter are, for the most part, unaware that such methods and devices even exist. The former are too hidebound to consider building alternatives. Progressive designers and contractors desire more unusual, avant-garde solutions to reinforced concrete construction than that offered by a mere 2-×-6-foot framework for an inexpensive panel. Con-

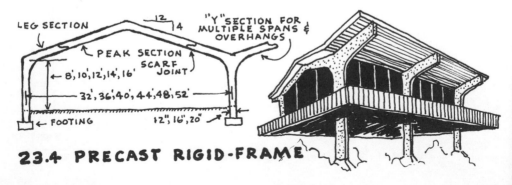

23.4 PRECAST RIGID-FRAME

crete systems, like the Youtz-Slick Lift Slab, in which floor and roof slabs for upper stories are poured on the ground level and are lifted into position by radio-controlled, hydraulic jacks, offer more challenge and mere prestige value to the high-price builder and his affluent clients.

Another popular structural technique, utilizing reinforced concrete, is to use prestressed beams and girders. Now becoming standard practice everywhere with the use of electric, hydraulic, or mechanical force to reinforce members, a beam is prestressed to withstand excessive compression loads. This technique may represent low cost for the mass builder, but the expensive, specialized equipment for this work is prohibitive for owner-builder use.

Architects too often conjure up techniques, such as large-scale lift-slab or prestressing projects in order to avoid the careful design, the painstaking thought, and the cultivation of sensitive imagination that accompany good architecture in its quest of housing solutions. In any event, such complicated, high-price solutions are of little interest to us here. They are anything but low-cost or simple for those interested in decentralized home construction.

24 WOOD

A general contractor who pays his carpenter the 1974 union-wage rate of about $9 per hour must, himself in turn, charge his client $18 for each hour that the carpenter works. The difference between these two figures is absorbed in the socialized benefits the contractor must pay for the worker, in the contractor's overhead and profit, and in wasted time. Consider, for a moment, the wasted time taken with bent nails, with pulling them, getting replacements, and beginning the same task over again. Multiply this probable number of poor-quality nails so mis-used by the 65,000 wire nails that it ordinarily takes to contribute to the support of a small house. There is a need for *quality* nails!

Take into account the one-third saving in the costs of stud lumber and its erection by spacing studs on 24-inch centers instead of on the code-enforced, 16-inch centers. In the 1950s, the University of Illinois Small Homes Council found that a 40-cent stud actually cost $5 by the time it was pounded into place. At today's 1974 prices, that 40-cent stud costs $1.50, or more than $3\frac{1}{2}$ times its 1950 cost. In-place costs of the stud today would, therefore, easily be $3\frac{1}{2}$ times more than the $1.50 price per stud of the 1950 priced lumber.

Professor Albert Dietz of MIT once stated that "practically all small homes built today use too many studs. You can't say they are over-engineered, because they aren't engineered at all. They are just over-built." From a standpoint of engineering and for the vertical loads imposed on most housing and for building spans of up to 32 feet, the 2-×-4-inch stud construction is adequate when the studs are spaced on 8-foot centers.

In-place nailing costs for wood-frame houses indicate that, on a weight basis alone, threaded nails can replace twice as many cut flooring nails. This revelation is without reference, either, to the superior performance of the threaded nail. The actual cost saving of 28 percent does not include corresponding labor savings. The permissible load for a single plain-shank nail, driven parallel to the wood grain, is from 80 to 90 pounds. A threaded nail of the same size will provide as much as ten times or 800 pounds of permissible load-holding power. Toothed-ring or split-ring connectors will provide twenty times or up to 1,900 pounds more holding power than that of a threaded nail.

All of the above discussion is by way of introducing into this chapter on wood the fact that wood-house-framing methods can be unduly complex, inefficient, and costly, or that they may be simple, effective, and inexpensive. The material, wood, can itself be a perfect choice for performing simple building tasks. Wood possesses high tensile and compressive strengths, and the high strength-to-weight ratio of some woods gives the structure a relative immunity to vibration. We certainly should not discount the possibilities of building with wood merely because it is so grossly misused in current building practices. But, before the owner-builder attempts to construct his house of wood, he would do well to attempt some understanding of the structural properties and the behavioral characteristics of this versatile material. Builders too often choose a high strength wood for siding when what is really called for to accomplish this aspect of one's building task is wood having good weathering properties or having good receptivity to paint or to another finish. Joists are often chosen without proper regard for their bending factor, whereas stiffness is more appropriate to joist selection. One's concern should be for joist dryness and for minimum shrinkage.

Wood can best be thought of as a reinforced plastic. Its primary constituents, cellulose and lignin, are also the main ingredients of commercial plastics. Lignin is the adhesive that gives strength and rigidity to wood. Cellulose is nature's strong material. It is made of long-chain molecules, which line up with the long axis of the tree. This explains why wood splits along the grain, but, by the same token, why it must be chopped or sawn across the grain. It also explains why, when it dries, wood shrinks much less in its length than it does in its width. Moisture in wood is not held inside the molecular chains, but rather between them. When evaporation occurs, the chains themselves contract very little, but do draw together in closer contact. Shrinkage is, therefore, lateral rather than longitudinal. While the tree is living, both the cells and the cell walls are filled with water, but, as soon as the tree is felled, water within the cells, called "free water," begins to evaporate. Shrinkage in the wood occurs only after the free water is fully evaporated.

Investigations at Virginia Polytechnic Institute have disclosed the fact that nail-holding power decreases by three-fourths when driven into green or only partially seasoned lumber. While seasoning, the wood shrinks away from a nail shank, reducing friction between the nail and the surrounding wood. There is also a deterioration of moist wood, a chemical digestion of the wood cellulose, due to the formation of iron oxides on a nail in the presence of wood acids. An unsheathed house frame assembled with green lumber loses during seasoning more than one-third of its initial resistance to racking.

Besides an increase in strength and in nail-holding power, air-dried or kiln-dried lumber holds paint and preservatives better and is less subject to attack from fungi and insects. It is fungi, in a moist atmosphere, that cause decay in wood. Since wood will not rot when dry, the term dry rot actually refers to the rot that leaves wood, when dry, in a brown, often crumbly condition that is suggestive of extreme dryness. Dry rot is caused by certain fungi which, by means of special water-conducting strands in their make-up, are able to carry water from the surrounding soil into a building.

Before purchasing any major amount of lumber, the owner-builder should make a rough check of its moisture content. Select a half-dozen flat or plain-sawn boards from a lumber pile and cut a sample from each. The sample should measure 1 inch along the grain and be cut to include the entire width of the board. It should be cut at least 1 foot from the end of the board. Trim the sample so that it will measure exactly 6 inches in width, and put it in a warm, dry place for 48 hours or longer. The 6-inch dimension is then measured to determine shrinkage. A shrinkage of over one-fourth of an inch shows the wood to be unsatisfactory for framing. A shrinkage of over $\frac{1}{8}$ inch means that the wood is unsatisfactory for interior trim or finish.

As many advantages as there are to the use of seasoned lumber, its use will not be widely practiced as long as the cost remains substantially higher than green lumber. A builder can maintain quality in construction and yet make substantial savings by buying less expensive, "green" lumber, by using threaded nails to erect framework, and by allowing the house shell to season prior to the application of interior finishing.

Results of an experiment done by Virginia Polytechnic Institute found that a house frame built of green lumber and assembled with plain-shank nails could withstand a racking load of 360 pounds. It failed when the nails pulled out of the end stud. In contrast, another frame of green lumber assembled with annularly grooved nails could withstand loads of 1,640 pounds and only failed when the base plate broke.

If the house frame is not sufficiently seasoned, nail popping will result after the installation of interior panels, for, in the seasoning process, wood shrinks as it dries out and it swells as it absorbs moisture. Nails, penetrating this alternately shrinking-and-expanding lumber, cannot appreciably change in length. This results in a backing-out of the nail. Some carpenters use longer nails to prevent nail popping, presuming that greater holding power comes with longer length or heavier gauged nails. On the contrary, nail popping has been found to be directly proportional to both lumber shrinkage and to the depth of nail-shank penetration. When buying nails, remember that the most effective nail is the shortest nail providing sufficient holding power, since it offers maximum resistance to nail popping. Elaborate house-framing studies by the Housing and Home Finance Agency have disclosed that nothing larger than the 3-inch long, tenpenny nail need be used in attaching two, 2-inch members in a simple lap splice.

Another possible solution to the popping problem would be the use of cement-coated, plain-shank nails. True, the withdrawal resistance of these special nails is improved by 75 percent as soon as they are driven, but the withdrawal resistance of cement-coated nails still decreases as much as 50 percent during the wood's seasoning.

From the standpoint of nail popping in green-lumber framing alone, I have come to feel that the owner-builder can ill afford the use of plain-shank, common wire nails in his wood-frame construction. For centuries, the joining of wood has been a bottleneck to the utilization of this material to its fullest value. In the past, for this reason, contact areas of the jointed members had to be large, and this fact governed member sizes more than did the actual working stresses on the lumber. But today we have synthetic resin-adhesives and hardened, threaded nails to create joints that are even stronger than most of the woods from which they are assembled. A special nail has been designed for every specific, wood-framing operation, whether it is cement-coated, cut, etched, barbed, knurled, twisted, fluted, or threaded.

Threaded nails have remarkable holding qualities, since wood fibers penetrate the grooves made by the threads, and this fish-hook-like action keeps the nails in place. The slender, grooved, hardened, high-carbon steel nail is more than one gauge smaller in diameter than the common wire nail. Being more slender, it can be driven faster and with less energy. It is also less likely to split the wood, so that these nails can be spaced closer together and nearer the edge or nearer the end of the wood.

About 1 percent of the cost of a wood house is for nails. If threaded nails, which have twice the withdrawal resistance, were used throughout the construction, the cost of a $5,000 house would be increased by $20. This additional expenditure would be more than offset by the resulting increased strength and building performance. The unsheathed framing would provide from four to six times greater lateral thrust resistance. The wood floor would not squeak, spring, cup, or buckle. Siding would not creep or pop. With threaded nails, two-and-a-half times greater holding power is given to asbestos shingles, 50 percent greater axial withdrawal resistance is given to asphalt roofing shingles, and 40 percent increased holding power is given to plaster board.

The problem of wood joining has been only partly solved by the development of improved nailing. Since World War II, two other improvements in fastening have contributed to making wood usable as a continuous material. The first improvement was the development of mechanical connectors, and the second improvement was the invention of strong glues.

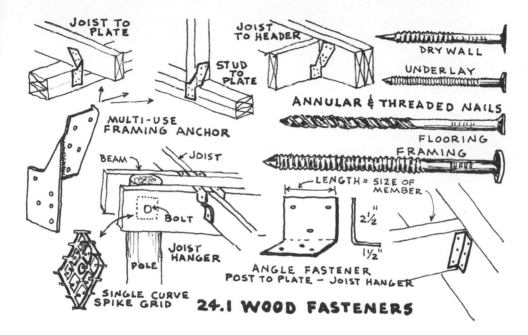

24.1 WOOD FASTENERS

Metal plates and gussets are on the market for use in strengthening wood joints, particularly against a direct pull. They are made of heavy steel and have specially designed teeth stamped into them. Held in place merely by nails, they make one nail do the work of several. Bent, predrilled, steel U-grip hangers can often be used to advantage to locate and to support 2-inch beams and joists. Framing anchors are also advantageously used at joints between plates and rafters and where studs attach to sills. Whereas toenailing prevents sidewise movement, anchors give protection against uplift.

Shear plates and tooth-ring and split-ring timber connectors are some of the best devices for use at joints that are fixed and permanently bolted. Toothed rings and shear plates are installed by forcing them into wood, so that half the depth of the connector is embedded in each of the two lapping members. Split-ring connectors are installed in precut grooves, with half of the ring being seated in each of the two, overlapping members. A bolt, passing through these members and through the center of the ring, completes this type of pin joint. Here in the West, we commonly use the Teco brand of fastener in the manner illustrated above.

TABLE 24.1

PERCENT INCREASE IN STRENGTH OF ROUGH LUMBER
OVER DRESSED LUMBER

Size	Bending Strength	Tensile and Compressive
2 × 4	50	36
2 × 6	40	31
2 × 8	40	31
2 × 10	36	30
2 × 12	34	28
1 × 4	36	42
1 × 6	46	37
1 × 8	46	37
1 × 10	42	35
1 × 12	40	34

If commercial bracing anchors are not available, or if the owner-builder prefers to save money by fabricating his own, a simple angle fastener can be made of bent, 16-gauge, galvanized sheet metal. One finds, by adapting the fastener length to fit the wood member to be fastened, that any number of anchor combinations can be made. For example, one can fabricate fasteners for joist to header, post to beam, stud to sill, and stud to plate.

The gluing of wood may be compared to the welding of metal for it makes a continuous member of several or more pieces. Gluing timbers is not a recent innovation, but, with the development of plastics, much more durable and waterproof glues have been produced. Common casein glue is still one of the best, least expensive, and more available of glues. It is especially suited to rough lumber, where joints cannot be made completely tight. Only water is added to the dry casein powder, but, once it has set, it is completely waterproof.

When gluing, use rough lumber in preference to dressed lumber. As well as being less expensive and more readily available from small, decentralized mills, rough lumber also contains a substantially greater strength factor, indicated in the chart above. Generally, rough lumber has better weathering characteristics. By dealing directly with a small milling operation, an owner-builder is more likely to secure, if he so insists, the very best cutting. This is the explanation: Large, commercial lumber mills prefer to slash cut a log. From a mechanical, sawing

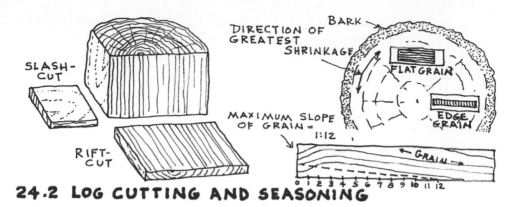

BARK
DIRECTION OF GREATEST SHRINKAGE
SLASH-CUT
FLAT GRAIN
EDGE GRAIN
MAXIMUM SLOPE OF GRAIN = 1:12
RIFT-CUT
GRAIN
0 1 2 3 4 5 6 7 8 9 10 11 12

24.2 LOG CUTTING AND SEASONING

point of view, this operation is faster, and there is less waste involved for the logger. However, a slash-cut board has less strength and offers less wear than a board which has been rift cut. Rift-cut hardwood is said to be "quartersawn," while rift-cut softwood is called "edge-grain" lumber. As illustrated, edge-grain (rift-cut) lumber shrinks and swells less in width. It also twists, splits, and cups less, as well as receiving paint or other applied finishes better. Where flat grain, slash-cut lumber is used structurally for its bending strength, the maximum slope of the grain should not exceed one in twelve. See the illustration above.

The correct selection of joining devices is as necessary to building assembly as the proper selection of lumber is prerequisite to successful framing and finishing. Being plentiful in many parts of the world and, therefore, relatively economical, wood is also easily worked—cut, drilled, or nailed—with simple tools. The choice among species, varieties, qualities, and grades of wood is sometimes staggering, and care must be taken to make a proper choice of the material for each specific aspect of the building operation. The following information is intended to assist one in making proper choices. (Classification is made relative only to the more important structural and behavioral characteristics.)

GENERAL CLASSIFICATION OF WOOD

Hardness in wood is the property that makes the surface difficult to dent, scratch, or cut. We look for hardness in wood for use in flooring and in furniture making, but we avoid this property where frequent cutting and nailing is required, since hardwood is more likely to split when nailed and is generally more difficult to handle, as well as being usually more expensive.

Weight is a reliable index of the structural properties of wood. That is, a heavy piece of dry wood is stronger than one lighter in weight.

Shrinking and swelling take place in wood as it alternately absorbs moisture and then dries out. About one-half of the shrinkage occurs through air drying and about two-thirds is removed through kiln drying. One fact that is especially important to know before selecting flooring boards is that an edge grain or quartersawn board will shrink about one-half as much at right angles to the annual rings of the lumber as it shrinks parallel to these rings in flat-grain lumber. Edge-grain wood of a species having a high shrinkage factor will prove as satisfactory for flooring use as flat grain wood of a species with an inherently lower shrinkage factor.

Warping of wood is closely allied with shrinkage. Lumber that is cross-grained or that is cut near the central core of the tree tends to warp when it shrinks. Quartersawn dry wood warps least.

Paint-holding depends upon a number of factors, such as the kind of paint selected and the circumstances for its application—as well as upon the type of wood used. But, in general, paint holds better on edge-grain wood than on flat-sawn boards. The bark side of flat-grain boards is more receptive to paint than is the pith side.

Nail-holding properties are usually greater in the denser and harder woods, but, as already mentioned, nails eventually lose about three-fourths of their full holding power when driven into wet wood. Blunt-pointed nails have less tendency to split wood than do sharp-pointed ones.

Decay in wood is caused by moisture and microbes. Wood is inherently vulnerable to moist and stagnant air, and it is susceptible to attack from fungi and insects. Moisture in wood is also directly responsible for plaster cracks, air leakage, pulling of fastenings, vibration of floors, peeling of paint, and the warping and sticking of doors and windows. It is a wise owner-builder who insists on dry lumber, especially for interior use. (For the exterior use of green lumber, consider again the previous comments in this chapter on its use.)

Bending strength is a measure of the horizontal load-carrying capacity of wood. Rafters, girders, and floor joists all require high bending strength. Viewing one's lumber selection cross-sectionally, a small increase in the height of a beam, or of a horizontal member, produces a greater increase in the bending strength than it does in the volume of the beam, whereas an increase in the width of a beam increases the volume and the bending strength of that beam proportionally. That is, an increase of 1 inch in the height of a 10-inch beam will increase its volume 10 percent, but the bending strength of the same beam is increased by a considerable 21 percent. An increase of 1 inch in the width of a beam, 10 inches wide, will increase both its volume and its bending strength by only 10 percent, or equally. Both the elastic limit and the ultimate strength of wood are higher under short-time loading than they are under long-time loading, so that wood is able to withstand considerable overloads for short periods or, conversely, smaller overloads for longer periods.

Stiffness is a measure of the resistance of wood to bending or to deflection under a load. A 10-inch joist has only about one-fourth more wood in it than an 8-inch joist, but placed on edge in a building it is more than twice as stiff. In studs, too, stiffness is more important than actual breaking strength, since it is deflection that must be reduced to a minimum in order to avoid plaster cracks in ceilings and vibrations in floors.

Toughness is a measure of the capacity of wood to withstand suddenly applied loads. Wood high in shock resistance withstands repeated shocks, jars, jolts, and blows, and gives more warning of failure than does non-tough wood, an important characteristic in beams and girders.

Wear resistance is higher in edge-grain wood than in flat-grain wood. This resistance is greater on the sap side of the wood than it is on the heart side. It is more evenly distributed in clear wood than it is in wood containing knots.

Knowing the physical properties of wood and the methods of attaching wood members is only the first or the preliminary-thought stage in building a house of wood. Subsequent to achieving this basic understanding, one must choose wisely some system of wood construction, and the structural system chosen depends, in turn, on a host of other factors—from general engineering principles affecting the design of roofs, walls, and floors, to such commonplace considerations as the availability of skilled labor and necessary equipment. These are factors that will be considered in the next two chapters.

TABLE 24.2

A GENERAL CLASSIFICATION OF WOODS

A = Among the woods *high* in the particular respect listed
B = *Intermediate*
C = *Low*

	Beech	Cherry	Maple	Oak	Cedar, incense	Cedar, eastern red	Cedar, western red	Cypress, southern	Fir, Douglas	Fir, white	Hemlock	Larch, western	Pine, north white	Pine, south yellow	Pine, ponderosa	Pine, sugar	Redwood	Spruce, eastern
hardness	A	A	A	A	C	A	C	B	B	C	B	A	C	A	C	C	B	C
freedom from shrinkage	C	B	C	C	A	A	A	B	B	B	B	B	A	B	B	A	A	B
weight (dry)	A	B	A	A	C	A	C	B	B	C	B	A	C	A	B	C	B	B
freedom from warping	C	A	B	B	A	A	A	B	B	B	B	B	A	B	A	A	A	A
paint holding	—	—	—	—	A	—	A	A	C	B	B	C	A	C	B	A	A	B
ease of working	C	C	C	C	A	B	A	B	C	B	B	C	A	C	A	A	B	B
nail holding	A	—	A	A	C	—	C	B	B	C	B	A	B	A	B	—	B	B
decay resistance of heartwood	C	—	C	B	A	A	A	A	B	C	C	B	B	B	B	B	A	C
bending strength	A	A	A	A	C	B	C	B	A	B	B	A	C	A	C	C	B	B
stiffness	A	A	A	A	C	C	C	B	A	B	B	A	C	A	C	C	B	B
strength as a post	B	A	A	B	C	A	B	B	A	B	B	A	C	A	C	C	A	B
toughness	A	B	A	A	C	B	C	B	B	C	B	B	C	B	C	C	B	B
number of knots	B	C	B	C	A	A	C	C	B	B	B	A	A	C	B	A	C	A
size of knots	B	B	B	A	C	C	A	B	B	B	B	C	C	A	B	B	A	C
proportion of heartwood	B	B	C	B	A	B	A	B	A	C	B	A	B	C	C	B	A	C

MAJOR USES (in order of importance)

- Beech: Flooring, veneer
- Cherry: Veneer, paneling
- Maple: Flooring, veneer, paneling, millwork
- Oak: Flooring, millwork
- Cedar, incense: Shingles, paneling, construction
- Cedar, eastern red: Flooring, millwork
- Cedar, western red: Shingles, siding, sash, millwork
- Cypress, southern: Siding, paneling, sash, millwork, flooring
- Fir, Douglas: Construction, sash, millwork, flooring
- Fir, white: Light construction, siding, sheathing
- Hemlock: Construction, sheathing, siding, flooring
- Larch, western: Construction, paneling, flooring, sash
- Pine, north white: Millwork, siding
- Pine, south yellow: Construction, sheathing
- Pine, ponderosa: Sash, millwork, light construction
- Pine, sugar: Sash, millwork, light construction
- Redwood: Sash, siding, paneling, millwork
- Spruce, eastern: Construction, millwork

References: U.S. Dept. of Agriculture Yearbook: 1949; Hornbostal: Materials for Architecture; Farmers Bulletin 1756

25 WOOD-FRAME STRUCTURES

Since the properties of wood and the proper techniques for its use only required the preceding chapter-long presentation, one cannot help wonder why it is that commercial home builders continue to neglect these more rational approaches to wood construction. Instead of heeding well-documented building methods and recent, demonstrable innovations alike, builders persistently choose to ignore a sensible, sound construction of the skeleton of the house (its framework) and prefer to go all out for affectation and trimmings. Frame houses built of conventional wall studding are overdesigned, while they yet remain weak at the joints. They are disappointingly inefficient, to say the least. And, both erection labor and materials of construction are wasted.

The structural engineering of houses has been studied extensively by various private and governmental agencies. One government report, *Strength of Houses,* maintains:

> Houses have never been designed like engineering structures.
> Since prehistoric times, safe house construction has been found by
> the tedious and wasteful method of trial and error. If the modern
> research that has proven so successful in the solution of other

problems had been applied to houses, not only would homes be more satisfactory as dwellings but, much more important, the cost would be much less. This would be an outstanding contribution to the problem of providing acceptable houses for the low-income groups in this country.

One cannot fail to note the lack of a rational, engineering approach to house-building practices today. It is customarily assumed that walls, flooring, and roofing contribute essentially nothing to the strength of a building. It is mistakenly supposed that all loads are carried independently by the frame. So, consequently, all stresses and loads are analyzed separately for each structural component. Design loads on walls, floor, and roof are calculated for compression, impact, and racking, yet, in these selective calculations, dead loads (the gravity weight of a building) are incorrectly added in a lump sum to live loads (objects or persons on floors of a building). To these misleading analyses the builder must add calculated wind, water, and snow loads. Then, to complicate matters more, the builder must be acquainted with the strength of various materials and with fastening methods, and with the degree of elasticity, stress, shear, and deflection for each independent member. The builder should also know the species of each variety of wood he uses, as well as the size of its effective cross-sectional area and the grade of the lumber to be used.

In a realistic engineering approach to the construction of wood-frame structures, the usual practice of analyzing each structural component separately is replaced by a consideration of the integrated structural effect of all components. A deliberate attempt is made to have the foundation and the roof function as extensions of the walls, eliminating the concept of a separate function between wall and roof, floor and foundation.

One of my favorite framing methods can certainly be included in the category of integrated, structural, house-building engineering. Called "California construction" and dating back to the gold rush days, the method eliminates the need for studding. Inasmuch as these early houses were built on stone piers with no studs, today's California building-code officials put this California construction on the renegade list. I have, however, recycled lumber from a number of these 100-year-old buildings and have wondered at how structurally sound and

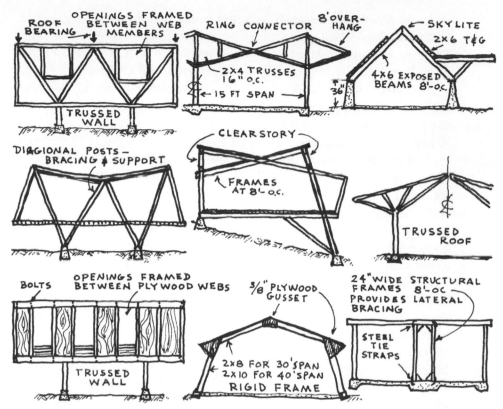

ROOF BEARING — OPENINGS FRAMED BETWEEN WEB MEMBERS
TRUSSED WALL

RING CONNECTOR — 8' OVER-HANG
2X4 TRUSSES 16" O.C.
15 FT SPAN
36"

SKYLITE
2X6 T&G
4X6 EXPOSED BEAMS 8'-O.C.

DIAGONAL POSTS — BRACING & SUPPORT

CLEAR STORY
FRAMES AT 8'- O.C.

TRUSSED ROOF

BOLTS — OPENINGS FRAMED BETWEEN PLYWOOD WEBS
TRUSSED WALL

5/8" PLYWOOD GUSSET
2X8 FOR 30' SPAN
2X10 FOR 40' SPAN
RIGID FRAME

24" WIDE STRUCTURAL FRAMES 8'-O.C. PROVIDES LATERAL BRACING
STEEL TIE STRAPS

25.1 RATIONAL FRAMING SYSTEMS

how well-preserved they remain. A detailed section on California construction is included below for the benefit of those readers who are able to or who care to circumvent unreasonable building codes.

Any structural system has four basic forces with which to contend. Compression and tension have been discussed previously in this text. Bending occurs when a weight or a force is placed on a member at a distance from its support, and shear is a thrust outward at right angles to an applied stress. With these basic structural concepts in mind, various framing systems can be evaluated and compared in relation to the strength per unit of material and to the time expended in the erection of the material. Wood was inefficiently used by early settlers when they built cabins of massive logs cut from trees. Later, with the advent of power-driven sawmills, wood-frame structures, having lighter members, were developed and less wood was required for a building. This post-and-girder structural system, see figure 25.2, provided a transition between the log-cabin building system and the currently employed vertical-stud-wall system of construction—which, in some ways, marks a decline from the effectiveness of the post-and-girder design.

The balloon-frame system of construction, which employs vertical studs in an old-fashioned box-type house, is still often used in two-story constructions. With ceiling joists supported on a ribbon board let into studs, it becomes possible to extend the studs, unbroken, to their full building height. Platform framing has almost entirely replaced balloon framing on single-story structures, especially in the West. This system is essentially a platform consisting of subflooring over joists, supporting a stud wall. The wall, in turn, carries the roof and ceiling construction. Platform framing lends itself well to panelized and tilt-up construction. In the tilt-up system, the wall may be framed on the ground to any desirable degree, and it is then tilted upward, braced, and fastened into place.

A major disadvantage of stud-wall construction is that it is vulnerable to fire. Since the rate of burning depends on the ratio of surface area to volume of timber, the use of many light-wood members makes stud construction fast burning. A wood structure that uses heavier and fewer pieces is safer from catastrophic fire.

The plank-and-beam structural system is becoming more popular with home builders. It has many advantages, including some resistance to fire. The National Lumber Manufacturers Association made a detailed study of the relative cost of plank-and-beam construction compared with conventional, joisted framing. The Association found that at least 25 percent of the cost of a building can be saved with plank-and-beam construction over the conventional system of building, since there are fewer pieces of lumber to handle and structural function is combined with surface finish.

Beams, covered with standard, 2-inch decking, may be spaced as far as 8 feet apart. A structural feature worth mentioning here is that, for the same, evenly distributed load, the plank that is continuous over two spans is nearly two and one-half times as stiff (rigid) as the plank that extends only over a single span.

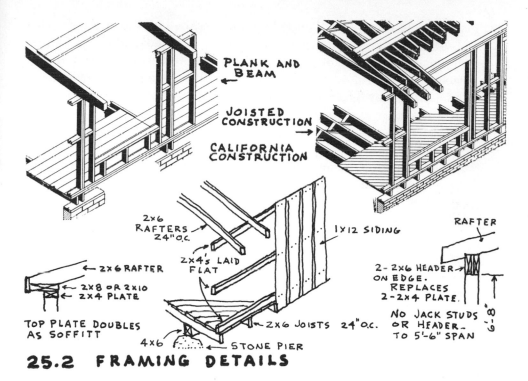

PLANK AND BEAM ←

JOISTED CONSTRUCTION →

CALIFORNIA CONSTRUCTION

2x6 RAFTERS 24"O.C.

← 2x6 RAFTER

← 2x8 OR 2x10
← 2x4 PLATE

TOP PLATE DOUBLES AS SOFFITT

2x4's LAID FLAT

4x6

1x12 SIDING

← 2x6 JOISTS 24"O.C.

← STONE PIER

RAFTER

2- 2x6 HEADER → ON EDGE. REPLACES 2-2x4 PLATE.

NO JACK STUDS OR HEADER_ TO 5'-6" SPAN

8'-0"
6'-8"

25.2 FRAMING DETAILS

Building researchers who have thoroughly studied wood framing methods claim that there are four structural innovations which, when employed, tend to provide a better structure with less cost. Therefore, prefabrication tends to decrease construction costs through a saving in time. Tilt-up erection permits more construction to be done at ground level, resulting in much greater efficiency. A continuous structural design (rigid frame) provides a structurally balanced and a self-supporting structure, which is stronger and at the same time requires less material. Finally, pole-frame construction decreases labor and material costs in such a variety of ways that a full chapter will be needed for a more thorough discussion of this method. Chapter twenty-six will describe the features of a pole-type wood structure that is tilt-up, prefabricated, and is built on the principle of a continuous, rigid frame of timber.

In the preassembly of wall units, the owner-builder may capitalize on some of the many money-saving building techniques developed by the highly competitive prefabricated home-building industry. The Lu-Re-Co panel system, for instance, is fully adaptable to owner-building and offers a 30 percent labor saving and a 10 percent framing saving over standard construction. Both interior and exterior panels, measuring 4 by 8 feet or less, depending upon window and door sizes, are framed in a special jig; each panel, including sheathing, siding, window, door, and even a prime coat of paint, is assembled on the ground.

Stressed-cover panels offer all the advantages of modern, prefabricated panel construction along with greatly improved structural qualities. This type of panel consists of a frame with a continuous plywood skin glue-nailed on each side of the frame. Loads are carried partly by the skin. Framing members can, therefore, be lighter and fewer in number than in standard framing construction. Strength and rigidity are both increased by this combination of interacting frame and plywood skin.

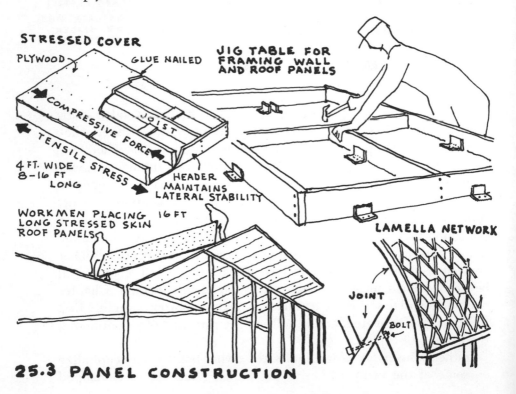

25.3 PANEL CONSTRUCTION

A stressed-cover panel is similar in structural action to a wide-flange steel beam. The top face of the beam carries compressive stress, and the bottom face carries tension. A joist acts as the steel web. High shearing stresses are transmitted throughout the plywood skin and framing members via glue-nailed joints. There are advantages to using plywood over ordinary laminated wood. The layers of plywood are placed with their grain running at right angles to each other, increasing strength and eliminating shrinkage.

The National Association of Home Builders' research house at Knoxville, Tennessee, uses thin, hardboard, stressed-cover panels on exterior and interior walls and on the floor, roof, and ceiling. All panels are 8 feet high, and they vary in width from 1 to 4 feet, making this system completely modular in planning and in details. If maximum benefit from panelized construction is to be realized, all components must conform to a module, which is a common unit of measurement. This will permit efficient assembly and eliminate all waste, fitting, and extra cutting operations on the job.

From the point of view of ultimate efficiency, the stressed-cover principle of construction demonstrates how unreasonable the frame-work-and-covering, studs-and-sheathing approach to construction really is. The skin-and-bones concept of structure is both wasteful and inefficient. Studs, rafters, and joists are first erected and then sheathing is applied as a cover. The sheathing used in this manner is a structural liability, a dead weight, a parasitic covering.

The greatest structural potential in wood construction today seems to lie in the construction of curved-skin enclosures. The principle of curved skin is based on the engineering concept that all material in a structure should contribute directly to its strength. It is a concept of integral structure, offering exciting prospects to any wood-oriented owner-builder.

The Lamella system, first developed in Germany in 1923, takes full advantage of the high strength of compression in wood used parallel to its grain. Loads are spread evenly over the entire network and are resisted by bolt-ties at each diamond-shaped diagonal. Lamella is essentially an arch composed of many short pieces of wood. Buckminster Fuller employs a small-component, triangulated system for the enclosure of space with his hemispherical domes. The Fuller dome develops extremely high strength-to-weight ratios and structural efficiency.

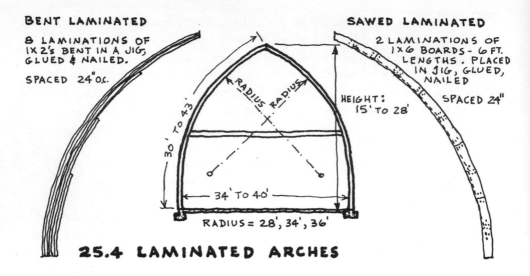

BENT LAMINATED

8 LAMINATIONS OF
1x2's BENT IN A JIG,
GLUED & NAILED.

SPACED 24"o.c.

SAWED LAMINATED

2 LAMINATIONS OF
1x6 BOARDS- 6 FT.
LENGTHS. PLACED
IN JIG, GLUED,
NAILED

SPACED 24"

RADIUS ÷ RADIUS

30' TO 43'

HEIGHT:
15' TO 28'

34' TO 40'

RADIUS = 28', 34', 36'

25.4 LAMINATED ARCHES

Short lengths of rough, home-sawn lumber can also be owner-laminated to create a Gothic-like arch. A 40-foot-wide, clear-span building having as much as 28-foot height can be made of short pieces of laminated 1- by 8-inch boards, as illustrated above. An alternative method for building laminated arches is the glue-nailing of 1 by 2s in a jig. As many as 8 layers, laid flat, can be glue-nailed.

Some builders have gone back to building log cabins and have developed a solid-bearing design with good results. Where cut lumber or other building materials are not readily available or where it would be too costly to import other building materials, one might consider building a log house. In general, however, log building is not a viable method of construction for the uninitiated to attempt. Working with logs is difficult, dangerous, and time-consuming work. No little amount of skill is required to properly join and to weatherproof logs. Where trees are the only available building-resource material, one might consider the advantages of investing in an Alaskan sawmill (Granberg Industries, Richmond, California). A simple attachment on a regular chain saw enables one to split small logs which are to be used vertically. It is on the order of the palisade of pioneer days. As illustrated below, staggered, split logs set vertically in a curved plan, provide a strong, well-insulated, weatherproof, and extremely low-cost wall structure.

A modern form of log-cabin construction uses 4- by 8-inch, tongue-

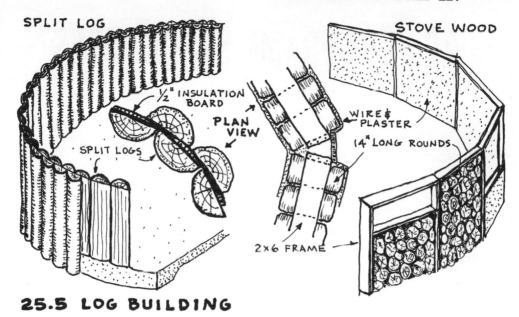

SPLIT LOG

STOVE WOOD

½" INSULATION BOARD

PLAN VIEW

SPLIT LOGS

WIRE & PLASTER

14" LONG ROUNDS

2×6 FRAME

25.5 LOG BUILDING

and-groove, sawn logs. Some methods of cabin construction use ½-inch steel rods, spaced every 4 feet and threaded through the logs. Another log manufacturer, National Log Construction Company of Thompson Falls, Montana, has patented a system for boring out the heart of the entire log. This air-lock cavity allows the log to season evenly, minimizing the cracking and checking of the log. The logs are also tongue-and-grooved on their exterior-seating edge, providing weatherproofing.

Early settlers in the Ottawa Valley discovered a method of building with wood which requires minimum labor and material cost. Dried cedar firewood logs, 14 inches long and averaging 8 inches in diameter, are bedded in cement in the manner that one stacks firewood neatly along a wall. The outside and inside of these walls are then cement plastered. A builder in Florida has constructed a number of stove-wood houses, using 7-inch-thick palm-tree sections. His four-room bungalow required only $28 worth of palm logs.

The range of possible systems for the construction of wooden structures, a range between cemented stove-wood and stress-skin arches and panels, gives the wood-oriented owner-builder a wide choice for his building interests.

26 POLE-FRAME STRUCTURES

The first of a small number of architect-designed pole-frame residences was built in California in the late 1950s. Even though the Doane Agricultural Service had, prior to this time, been able to popularize pole-frame barn construction for the Midwestern farmer, it was a rare builder then who could imagine how extensively poles would be used for homes today. If there exists just one symbol of contemporary American life it is the ubiquitous telephone pole, and it is, perhaps, fitting that an equally viable use in alternative housing has been found for this timber. It is interesting to note the disproportionate number of pole-frame houses which now appear in the contemporary work *Shelter*, Lloyd Kahn's recently published selection of self-built homes. It was Kahn who popularized dome construction in his *Domebook One* and *Domebook Two*, but it now appears, as Kahn seems to indicate, that a shift from dome construction to pole construction has taken place, and, in my judgment this denotes one giant leap for mankind.

Pole-frame building represents, along with prefabricated, tilt-up construction and rigid, wood-frame structuring, the very best thinking available to wood-oriented owner-builders today. Originally, pole construction was developed to give farmers a method of building which was, at once, much more economical than methods previously known to them as well as a method that would provide much needed wind resistance for their buildings. It also made possible long spans for the creation of large areas to accommodate hay lofts and storage. The subject of this chapter, therefore, will be how this was accomplished for the American farmer of yesteryear and how the owner-builder of today can benefit from the use of poles for building.

The difference between pole framing and stud-wall framing is the fact that poles, if properly pressure treated, can be embedded deep into the ground, providing a total bracing effect against the force of strong winds. The poles, extending high above ground level, provide support for floor and roof timbers alike in a continuous framework—from 5 feet below ground level to the very peak of the roof. Walls are hung from the poles supported by a system of horizontal girts. It is the point where the building proper joins the foundation poles;

26.1 POLE FRAME HOUSE

that is, where the stud wall functions in hinge fashion, that this building method exemplifies the principle of continuous structure.

The hypothetical house in Figure 26.1 illustrates some of the essential features of pole building. Different architects, of course, interpret pole building in different ways, using materials in terms of how they may best coalesce. I have evolved my own favorite concepts of pole construction, devised in part from personal building experience and, to a large extent, from engineering research. Special indebtedness must be extended to Homer Hurst of the College of Architecture, Environmental Systems Laboratories, Virginia Polytechnic Institute, for what is, perhaps, the best, most innovative thinking on pole structure today.

In reference again to the above-illustrated, hypothetical pole-frame house, notice first its siting. A pole-frame house is readily adaptable to difficult site conditions, whether it is a steep, rocky, heavily forested, or otherwise relatively inaccessible site or whether it is one's intent to minimize site disturbance and site preparation. A design of two or more levels seems called for when poles are used, for, unlike dressed lumber, poles are available in considerable length at very reasonable cost. The author's design shows a lower car-parking level and a sleeping loft-deck level above an intermediate, group-living level.

Cantilevered decks and room extensions can be a very practical bonus for pole-building owners. The very nature of heavy, timbered beams, paired on either side of the pole support, suggests its use in a cantilevered effect. Frank Lloyd Wright was one of the first notables to promote the cantilever principle of free space in small-home building. Nowhere is the principle more applicable than it is in building with poles. Cantilevered beams make possible greater spans for the same size timber, since the load of the cantilevered section of the beam counteracts the load on the center span. Stresses on the center span are thereby reduced, making possible increased spans or increased loads. These cogent structural principles appear to be too elementary for building-code officials to appreciate or to accept, and so it is that this excellent building method remains unavailable to the house-needy public in code-enforced areas.

Pole building suggests a natural, rough-hewn aesthetic. Poles should be exposed on the outside of the building, with unfinished, vertical boards nailed between these upright supports. A round pole is about 20 percent stronger than a squared timber, and the building design should articulate this substantial vigor. Incidentally, the reason that round poles are stronger than squared timber is that wood fibers in a pole grow around knots and defects, blending them into the strength of the total mass of wood.

When building a pole house, an owner-builder makes better use of a chain saw than an electric hand saw. This building method is also more applicable to the use of unskilled labor, for the builder handles fewer (though heavier) timbers, and less precision cutting of members is necessary. Much of the cutting and bracing of many small pieces, common to conventional construction, has been eliminated. Light frame structures are more susceptible to fire hazard, while the heavier, more widely spaced support members in this kind of construction lessen the prospects of a consuming fire destruction of one's work and shelter.

One feature of pole building, apropos of owner-builder attention, is its adaptability to owner-builder design. The principle of pole-frame structure is simple and can be readily grasped by most novitiates. Spacings of the poles can be as much as 14 feet apart, with 20-foot-beam spans, which makes a modular design wholly feasible. Future bays can also be readily added to the building, expanding the floor area as needed.

A few structural factors must be understood, however, before the owner-builder can confidently go forward with the design and the construction of his pole house. Perhaps the most important factor involves the embedding of the pole itself. One needs to determine, first, the type of soil on his site, and, equally crucial in cold country, to know frost depth. It is below this depth that the bottom of the pole must be set to prevent the pole from being displaced when the moisture content of the soil freezes.

The quality of foundation soil has been rated from excellent to poor in Figure 26.2. The depth and width of the hole that is to be dug to receive the pole can be generally determined from this table. Keep in mind the fact that poles are expected to perform the dual function of vertical structure support and of bracing for the building. An 8-inch-thick footing of concrete, poured at the bottom of the foundation hole, contributes greatly to bearing and stability. The skin friction of back-filled earth against the sides of the pole-in-place also carries a major portion of the vertical load of the pole. Some builders prefer to fill concrete around the set pole. Experience has shown the

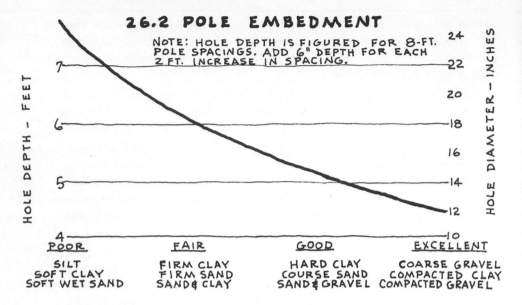

26.2 POLE EMBEDMENT

NOTE: HOLE DEPTH IS FIGURED FOR 8-FT. POLE SPACINGS. ADD 6" DEPTH FOR EACH 2 FT. INCREASE IN SPACING.

HOLE DEPTH – FEET

HOLE DIAMETER – INCHES

POOR | FAIR | GOOD | EXCELLENT

SILT
SOFT CLAY
SOFT WET SAND

FIRM CLAY
FIRM SAND
SAND & CLAY

HARD CLAY
COURSE SAND
SAND & GRAVEL

COARSE GRAVEL
COMPACTED CLAY
COMPACTED GRAVEL

author that a better practice is to tamp around the pole a granular (moist and sandy or gravelly) fill to promote drainage. When wet concrete is poured against a wooden pole, the pole swells before the concrete sets. When the pole dries it contracts and loosens, invalidating this entire procedure. Only enough back filling should be done to stabilize a newly set pole, for, when the floor and roof beams are secure in place and when the building is squared and made plumb, the poles can then be further anchored by thoroughly tamping soil around them.

The owner-builder should insist that the poles he buys be debarked, have a 6-inch top diameter, and be pressure treated. Before they are purchased, the poles should be inspected for their straightness and for adequate preservative treatment. If the preservative does not fully impregnate the wood, the pole's treatment will look spotty. Adequately preserved pine or fir poles will give as much as 50 years' in-ground structural service before decay sets in.

There is a good argument for using pressure-treated poles only for that portion of the structure that is buried in the ground. A less expensive, untreated frame can be erected from the platform floor upward, so the argument goes. Professor Hurst fully explored the problems associated with building rigid frame, prefabricated wall and

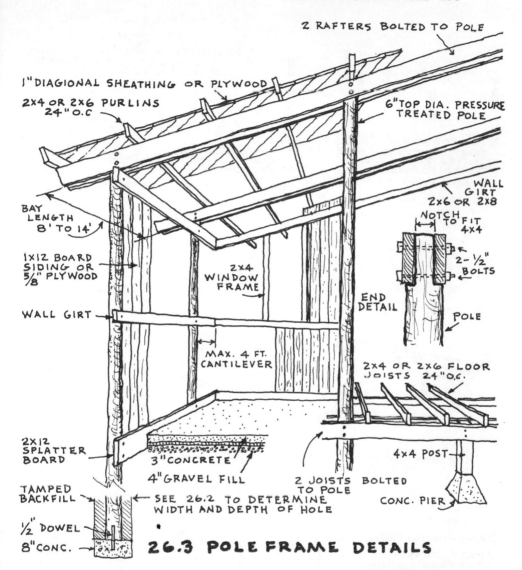

2 RAFTERS BOLTED TO POLE

1" DIAGIONAL SHEATHING OR PLYWOOD

2X4 OR 2X6 PURLINS 24" O.C

6" TOP DIA. PRESSURE TREATED POLE

WALL GIRT 2X6 OR 2X8

BAY LENGTH 8' TO 14'

NOTCH TO FIT 4X4

1X12 BOARD SIDING OR ⅝" PLYWOOD

2X4 WINDOW FRAME

2-½" BOLTS

END DETAIL

WALL GIRT

POLE

MAX. 4 FT. CANTILEVER

2X4 OR 2X6 FLOOR JOISTS 24" O.C.

2X12 SPLATTER BOARD

3" CONCRETE

4X4 POST

4" GRAVEL FILL

2 JOISTS BOLTED TO POLE

CONC. PIER

TAMPED BACKFILL

SEE 26.2 TO DETERMINE WIDTH AND DEPTH OF HOLE

½" DOWEL

8" CONC.

26.3 POLE FRAME DETAILS

roof units which were tilted up from a hinged construction secured to pressure-treated, foundation poles. The professor developed a self-supporting, structurally balanced system of housing construction, which is certainly applicable to owner-builder use. His buildings were determined to be 50 percent stronger than conventionally built struc-

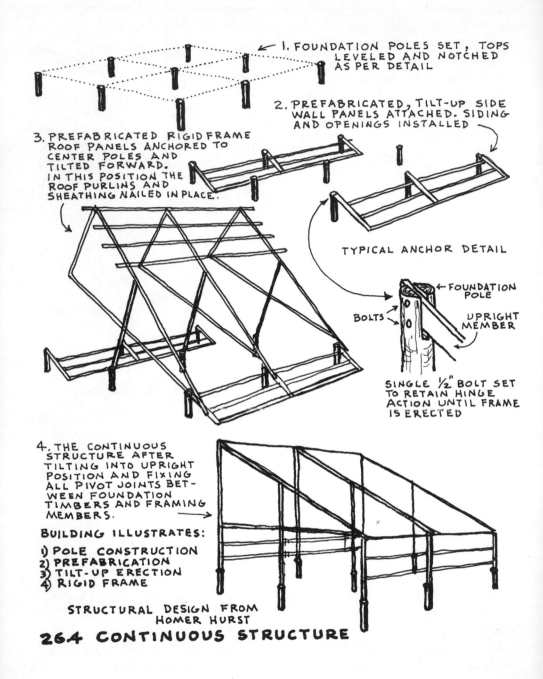

1. FOUNDATION POLES SET, TOPS LEVELED AND NOTCHED AS PER DETAIL

2. PREFABRICATED, TILT-UP SIDE WALL PANELS ATTACHED. SIDING AND OPENINGS INSTALLED

3. PREFABRICATED RIGID FRAME ROOF PANELS ANCHORED TO CENTER POLES AND TILTED FORWARD. IN THIS POSITION THE ROOF PURLINS AND SHEATHING NAILED IN PLACE.

TYPICAL ANCHOR DETAIL

← FOUNDATION POLE

BOLTS

UPRIGHT MEMBER

SINGLE ½" BOLT SET TO RETAIN HINGE ACTION UNTIL FRAME IS ERECTED

4. THE CONTINUOUS STRUCTURE AFTER TILTING INTO UPRIGHT POSITION AND FIXING ALL PIVOT JOINTS BETWEEN FOUNDATION TIMBERS AND FRAMING MEMBERS.

BUILDING ILLUSTRATES:
1) POLE CONSTRUCTION
2) PREFABRICATION
3) TILT-UP ERECTION
4) RIGID FRAME

STRUCTURAL DESIGN FROM HOMER HURST

264 CONTINUOUS STRUCTURE

tures. In one 1962 experiment three, open-type, clear-span buildings, suitable for agricultural and for light commercial use, were erected on the Virginia Polytechnic Institute campus. Two of the structures totaled more than 5,000 square feet and were erected with material and labor costs of 39 cents per square foot. This was less than one-half the total cost for a comparable stud-frame building. The author illustrates in Figure 26.4 his modification of the Hurst continuous-structure building system for the purpose of making it more applicable to owner-builder home construction.

27 COMPOSITE MATERIALS

An attempt has been made in the previous ten chapters to explore the full range of possible building materials. Some superior but less expensive methods of construction and less expensive wall-building materials have been suggested to the reader. Emphasis has been placed on the fact that present-day building practices have failed to provide for the low-cost housing needs of the majority of home builders. Virtually none of the commercially available building materials are applicable to truly low-cost building. Furthermore, if an owner-builder expects consistent, high quality, economical results, he must be sufficiently aware of current house-building alternatives and be free to consider completely new forms of structure, unorthodox methods of construction, and materials heretofore unheard of.

No attempt will be made here to list these new, unorthodox, or unheard-of building materials and methods used by man, but the utilization of waste and of readily available materials will be expounded as well as the principle of making composite materials. It is this principle of synthesizing various materials for their qualities of insulation, structural stability, workability, low cost, and so on, and not the particular formulas for various materials that must be understood by the owner-builder.

236

Practically any readily available waste material can be processed for composite filler. For example, in the Midwest section of our country, ground corn cobs have been successfully used as a filler for concrete. In the South, rice hulls have proved advantageous as a concrete filler. In the Northwest, sawdust has long been used as a concrete filler. The use of these and other materials will be discussed below to demonstrate the practical value of composite construction.

Cobcrete (corn cob) samples made and tested at a Michigan agricultural college were found to have a relatively low thermal conductivity of 3.0 k and a sufficient compressive strength of 1,000 lbs. p.s.i., which are average building requirements. Generally, a minimum of no less than 600 lbs. p.s.i. compressive strength is required. Either cast-in-place or precast blocks of this material can easily be fabricated. Fresh, dry cobs are first ground into pellet form by a farm hammer mill. After screening to the desired size of $\frac{1}{8}$ to $\frac{1}{4}$ inch, the cob pellets are placed in burlap bags and soaked in water for at least 6 hours. Following a 4-hour draining period, cobcrete is made using a mixture in these proportions: 1 part cement, 2 parts sand, 3 parts cob pellets, and $\frac{1}{4}$ part lime.

The first cement-and-rice-hulls block house built in this country in 1923 at Payne, Louisiana, still stands in excellent condition. Experiments at Louisiana State University have proven that a cement-and-rice-hulls composition has sufficient strength under compression and tension to meet ordinary, structural requirements. Furthermore, cement-and-rice-hull blocks, being lightweight, have high insulating properties. Good weathering properties also prevent the expansion, contraction, and subsequent excessive cracking of exposed blocks. A proportion of 1 part cement, 4 parts rice-hull ash and 2 parts rice-hulls was first used to make these blocks, but, later in 1953, it was found that the addition of clay gave still better results. Then, instead of cement, emulsified asphalt was used as a waterproofing agent in a soil and rice-hull ash mixture. The proportion used was 1 cubic foot of soil, containing not more than 85 percent nor less than 40 percent clay, 1 cubic foot of rice-hull ash, $\frac{3}{4}$ gallon emulsified asphalt, and $2\frac{1}{2}$ gallons of water. One-half of the total amount of required water and emulsified asphalt are first placed in a mixer and allowed to agitate for three minutes. The soil is then added, followed by the rice-hull ash and the remaining water. The average density of this composite is 70 pounds per cubic foot, which compares very favorably with adobe which is 100 to 120 pounds per cubic foot or with concrete which is 115 to 150 pounds per cubic foot.

Research into the use of a sawdust-cement composite for building construction started in 1930 at Oregon State College. Extensive work with this composite was also conducted in England prior to World War II. The principle advantages of the use of sawdust cement are its light weight, low cost, high insulation value, nail-holding capacity, and its resistance to freezing, burning, and termites. A major disadvantage of using sawdust cement is encountered as a result of its high shrinkage and expansion. Any composite material having cement as a binder undergoes a certain amount of shrinkage and expansion as a result of drying and rewetting. Sawdust undergoes an even greater volume fluctuation in its wet-dry cycle, contributing even more to the deterioration of the material by movement and subsequent cracking. By using a higher ratio of cement in the sawdust-cement composite, movement can be reduced. Careful, long-term curing of the finished panel is important. British researchers have shown in their work that, if smaller size sawdust-cement panels are coated with weatherproofing and are allowed to expand and to contract at some point of connection, the expansion-contraction damage is negligible. If sawdust is presoaked, it requires less water and less cement for the desired result.

Large, coarse, firm sawdust particles screened through a $\frac{1}{4}$-inch sieve proved to be the best sawdust-cement composite. In general, conifers make better sawdust-cement than do deciduous trees. Recommended species are spruce, Norway pine, jack pine, and aspen. Inferior species of softwoods are Douglas fir, red cedar, and Ponderosa pine. Woods having large amounts of organic matter, that is, gums, sugars, starches, and tannins, make poor quality sawdust cement. The high tannin-acid content of red oak, for instance, puts oak very low in the range of acceptable woods. British researchers found that by adding $\frac{1}{3}$ volume hydrated lime per 1 volume of cement to an otherwise unacceptable species of wood sawdust, tolerable results were achieved. An admixture of 1 percent diatomaceous earth also resulted in a far superior product. A proportion of 1 part cement, 1 part sand, 3 parts sawdust, and $\frac{1}{2}$ part clay or lime has proved to be satisfactory for either block or poured-in-place walls. The drying shrinkage of a mix of 1 part cement to 3 parts sawdust was found by the British to be .55 percent. By substituting half of the sawdust with sand, the shrinkage was reduced to .14 percent. Reducing cement content improved insulation value and reduced building cost. Two inches of the above mix has an insulating value approximately equal to that of ½ inch of standard fiber insulating board.

An interesting use for sawdust as a filler was developed not long ago by Carl Fabritz of Germany. His use of a formula that uses readily accessible raw materials makes a product with a lightweight, granular composition. A proportion of sawdust and cement is mixed, and, before its agitation, there is added a froth-forming agent (soapy water) with a small proportion of waterglass. A multicellular composition is produced as moisture is withdrawn from the mixture by the frothing action. In fact, so much is withdrawn that the walls of the froth globules are not supported by the binder (cement). As a result, the binder coats the grains of sawdust. The waterglass functions to support the froth-globule walls, holding the grains in suspension until all moisture is evaporated. When evaporation is complete, the binder sets and the froth globules gradually burst. The result is a composite product that is low-cost, lightweight, strong, and has good insulating properties.

Fabritz's technique for producing a lightweight-concrete product is only one of a number of air-entraining processes developed over the past forty years. Air entraining was first employed to improve the resistance of concrete to weathering, especially to freezing and thawing conditions. The cellular composition of air-entraining concrete creates a greater resistance to the capillary passage of water up through the concrete. Minute air bubbles act as a sponge, providing spaces where factors generally causing concrete disintegration can be dissipated. Air-entrained concrete is not significantly lower in strength, less water and cement are needed for mixing it, and its workability is improved. The major advantage of air-entrained concrete is, however, that it is lightweight and has a corresponding high insulation value.

Some air-entraining agents require special mixers, as, for instance, the English "Aerocem" and John Rice's "Bubblestone." The Rice apparatus for making a cellular-concrete product was first patented in 1937. It consists of a series of perforated cylinders, one within another, which revolve about a horizontal axis inside a cylindrical tank.

As the perforated inner cylinders revolve, air is trapped by a froth composition of phenols and aldehydes dissolved in water. Vinsol resin is another commonly used air-entraining agent.

A somewhat less involved procedure for achieving cellular concrete is to add a suitable lightweight aggregate directly to the mixture. There are a number of such aggregates on the market, and natural deposits of them can be found in various parts of the countryside. Several years ago a Texan built a complete house—walls, roof, and floor—with diatomaceous earth as the sole aggregate. Diatomaceous earth is formed from the decayed skeletons of tiny marine life. Deposits are scattered over wide sections of this country. In 1952, an experimental building using this aggregate was built at the University of Idaho. The 24- by 48-foot structure cost $2.85 a square foot, as contrasted with then prevalent costs of $5 to $8 a square foot using conventional materials. Proportions were 1 part cement, $1\frac{1}{2}$ parts diatomaceous earth, and 6 to 8 parts wood shavings from a planer mill.

In the eastern part of the United States, expanded slag as well as vermiculite is available for lightweight-concrete aggregate. In the West, pumice is perhaps the most popular and the least expensive of all lightweight aggregates. Scoria and perlite, also of volcanic-glass composition, are readily available. One major advantage of a lightweight aggregate is in the reduction of structural weight. Its use may reduce the dead load of concrete by more than one-third. Also, the thermal insulation, as previously stated, and the acoustical insulation values of lightweight aggregate are many times greater than those of standard concrete.

Reference was made in a previous chapter to the poured stone-rubble technique developed by architect Ernest Flagg. Another contribution made by Flagg to owner-built housing was a way of building nonbearing partition walls of wire lath and plaster. His system is most interesting. Two plasterers, one on each side of a single layer of wire lath, stretched from floor to ceiling, trowel against each other. After the first application, the wall is sufficiently strong for a finish layer to be applied without back support. The actual strength of reinforced plaster is astonishingly high. It is not unusual to find a plastered wall holding up a roof after the rest of the wall has been destroyed by fire.

Actually, in 1930, Major de W. Walker of Ireland found that plastered fiber concrete will support loads of 600 pounds per square foot.

On this basis, he developed a new method of strengthening concrete, called "No Fango." The process is simple. Fiber reinforcement is spread lightly over a framework of wood, reinforced concrete, or steel, and the fibrous material is then allowed to shrink. When shrunken, it is thoroughly impregnated with common cement slurry and it is concrete plastered. The finished wall is 1 inch thick. For outside walls, an air space is provided between two 1-inch layers.

The same idea of starched concrete inspired some experiments by Bernard Maybeck, renowned San Francisco architect, when building his own home. A framework was set up and wires were stretched 18 inches apart around the building. Burlap sacks were soaked in a very thin mixture of Portland cement, of powdered, lightweight aggregate, and of water. The soaked bags were then hung on the wires, like clothes on a line, to form a sturdy, low-cost, fireproof wall.

Obviously, the very nature of fiber-concrete construction demands equally unique structural design forms and practices. In the early 1950s, Dr. Kurt Billig, the very able director of building research at Central Building Research Institute, New Delhi, India, evolved a totally new building form using starched concrete. A house he displayed at the 1954 International Exhibition on Low-Cost Housing was a simple corrugated shell. Here, for the first time, a house was developed for that 90 percent of the population of India that cannot afford to pay for even a minimum-priced conventional structure. The CBRI shell house contained one-sixth the volume of building materials that would have been required for a conventional brick house of the same amount of floor area, and it cost about one-sixth as much to build. Where 313 man days would be required to build the conventional brick house, only 118 such man days were required to build the corrugated shell house.

The first step in erecting a corrugated shell structure is installing the falsework. This usually consists of tubular steel or wood-truss ribs, placed at from 3-foot to 8-foot intervals, depending upon the span of the arch. This formwork can be kept light in weight. Between each rib, a vegetable fabric is laid, such as hessian, jute, coir, sisal, or burlap. It was found that, whether stresses are due to static, dynamic, or thermal loads, vegetable fabric has the remarkable property of resisting high tensile stresses in all directions. Dr. Billig says:

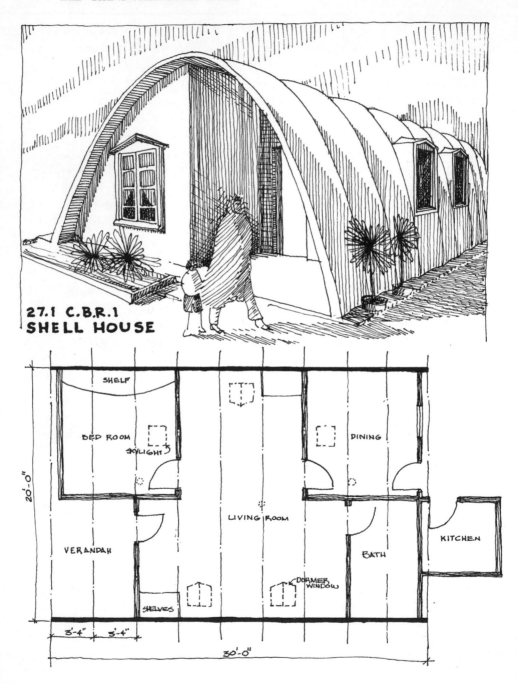

27.1 C.B.R.1
SHELL HOUSE

The dry hessian is stretched by hand as tightly as possible over the rigid falsework and securely fastened to it. By shrinking the fabric with water the fibres are stretched still tighter and in this state they are covered with a cement slurry by a brush. The cement is carried into the pores of the material where it remains. It maintains the shrinkage of the material, and initial tension in the jute is maintained as permanent tension after the concrete has set and hardened.

Absolutely no metal reinforcement need be used in corrugated-concrete-shell roofs, foundations, or floors. The shell is less than 2 inches thick, with possible spans of up to 60 feet. A day after the last plaster coat is applied, the falsework can be removed, and the structure can be expected to withstand 85-mile-per-hour winds.

In excessively hot or cold regions, it may be desirable to build double-skin shells. This simply involves placing another layer of fabric over mortar fillets, or over a layer of bricks, and then plastering as before. This cavity-air space improves insulating qualities considerably with a U factor of 0.37. In the tropics, it was noted that, as the air of the intershell cavity became warmed from the day's heat outside, it rose along the intrados of the arch, collected more heat in the upper regions of the shell, and then escaped through vents placed along the crown. The result was an effective natural-ventilation and efficient heat-insulation system.

Like Billig of India, Jack Bays of Cedaredge, Colorado, is an old-timer in the field of low-cost, composite building. For 25 years he directed an experimental-housing laboratory in Oklahoma City. His many building-material formulas are interesting and offer the owner-builder the epitome in low-cost building.

Bays's material consists of 12 quarts of asphalt emulsion and 8 pounds of shredded paper or cardboard mixed in equal parts with 12 gallons of a clay and water combination. The finished product, called Rub-R-Slate, can be troweled or sprayed on walls and floors. His favorite Rub-R-Slate construction consists of wall studs spaced on 30-inch centers with chicken wire tacked to both sides of the studs. The wall center, between wire coverings, is stuffed with straw or hammer-milled fiber to form a back-up for the Rub-R-Slate and to act as dead-air insulation. The Rub-R-Slate composition is plastered on both the inside and outside walls.

A two-bedroom home in Oklahoma was built for $1,200 with Bays's material. The walls consisted of used quart oil cans, stabilized by horizontal wires stretched at two-foot intervals and secured by a corner angle iron. These walls were further firmed and covered with a claylike sand, asphalt emulsion, and water mixture. The Rub-R-Slate coating was applied both inside and outside for a finish coat.

Jack Bays hit upon an ideal composite when he combined clay with emulsified asphalt and cardboard fiber. Clay is an excellent binder. Emulsified asphalt, of the slow-breaking type and commonly called "bitumul," provides the waterproofing. Cardboard makes up the fiber and the mass, contributing lightness and insulating qualities. Except for the bitumul, which is itself inexpensive, the materials can be gathered for the cost of hauling. Sample wall panels of Rub-R-Slate that I have built revealed the truly exciting qualities of this material. Even the act of plastering this composite onto chicken wire showed the material to be extremely responsive. To work with plaster is satisfying, but to work with Rub-R-Slate is pure joy.

The utilization of waste iron and tin cans for concrete reinforcement dates back to at least 1927, when George Watson of England developed and patented a process of flattening and shredding waste iron. To the budget-minded owner-builder, every city dump or local junkyard can be a source of concrete reinforcement by providing bed frames, iron pipe, wire fencing, wire cable, and so on, as well as the ubiquitous tin can.

When integral reinforcement is used in cement, even if it is nothing more than shredded tin, greater stresses can be withstood by the concrete. Expensive iron rods are not always necessary. In some countries, the scarcity of iron or a desire to use only available, natural materials necessitates the exploration of satisfactory substitutes for commercial iron rods.

One of the first experiments with the use of wood for concrete reinforcement was in 1914 by H. K. Chow, a Chinese student at the Massachusetts Institute of Technology. In Nanking, China, a few years later, a stretch of highway was laid down with bamboo as reinforcement. This successful experiment encouraged the Chinese to extend the use of bamboo to concrete reinforcement.

Then, in Italy in 1935, as a result of iron shortages due to war preparations, a number of wood-reinforcement experiments were carried on, and bamboo was found to be the most suitable wood for strengthening concrete. There is no chemical reaction between bamboo and cement mortar. The bond with concrete is much greater for bamboo than it is for any other wood, and bamboo has great tensile strength—about the same as concrete itself.

The Italian experiments showed that the modulus of elasticity in bamboo is about one-tenth that of steel. In other words, the cross-sectional area of bamboo should be ten times that of steel to give the same results. Other researchers, doing engineering research at Clemson, South Carolina, have pointed out that bamboo reinforcement increases the load-carrying capacity of a beam to four or five times that of an unreinforced member.

Even more significant than structural similarities is the cost differential between bamboo and steel. In tropical countries, bamboo cane will grow as much as 3 feet in one day. It grows without any special attention, and no particular problems complicate its harvest.

Clemson research reports that greater loads are possible if the bamboo is unseasoned, split, and treated with asphalt emulsion. The Italian experiments suggest a white-lead varnish treatment and a thicker protective coating. Concrete has excellent resistance to compression, but its modulus of elasticity is negligible. Bamboo can be economically substituted for steel to achieve this tensile strength.

In areas where bamboo is not available, timber can be employed to achieve similar results. Composite timber-concrete beams were built and tested in 1940 at the Talbot Laboratory, University of Illinois. The timber portion of the beams was designed to withstand tensile forces and the concrete portion was designed to provide compressive rigidity. The Talbot investigators found that the major obstacle to such a structural system is the high, horizontal shearing stress that exists at the junction between the two materials. Some mechanical device is needed to key the two parts of the beam together. Of the wide variety of shear connections examined, it was found that triangular, steel-plate units combined with iron spikes appeared to be the most satisfactory for producing integral beam connection.

Wood 2 × 4s can be used to reinforce simple, concrete slabs. Such a wood member, placed on the underneath side of a slab being poured, provides maximum tensile reinforcement, and such a member doubles as a means of attaching insulation and finish boards to either wall or

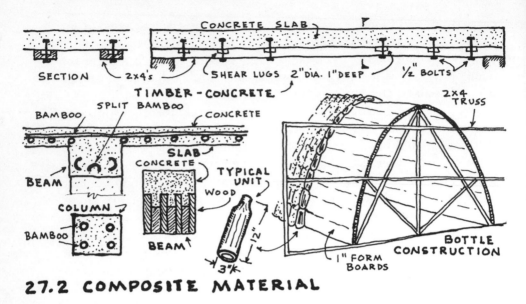

CONCRETE SLAB

SECTION — 2x4's — SHEAR LUGS — 2"DIA. 1"DEEP — ½" BOLTS

TIMBER - CONCRETE

SPLIT BAMBOO — CONCRETE

BAMBOO — SLAB — 2x4 TRUSS

BEAM — CONCRETE — TYPICAL UNIT — WOOD

COLUMN

BAMBOO — BEAM — 12" — 3¼" — 1" FORM BOARDS — BOTTLE CONSTRUCTION

27.2 COMPOSITE MATERIAL

roof slabs. These members can additionally be used to replace part of the form lumber.

Early Christians often employed unusual but effective building systems, which have since been little used or understood. Around A.D. 300, the church of San Vitale was built in Ravenna, Italy, with a dome constructed entirely of hollow earthenware. The urns were so formed that the end of one fitted into the mouth of the next.

This pot or bottle technique of construction had truly become a lost art until recently revived by French architect Jacques Couelle. Couelle's three-layer bottle vault was designed to span 49 feet without using reinforcing steel. He spanned 26 feet on a flat ceiling with this method of construction. The neck of one bottle receives the open end of another, with ordinary cement securing the union between the two bottles. Longitudinal burrs along the sides of the bottles further bond the union of bottles and cement. Each bottle is finally encased in a thin layer of cement.

Couelle reports a saving of 50 percent in cement and a saving of 30 percent in weight over conventional, concrete-block construction. M. Ros of the Swiss Laboratory for the Testing of Materials claims that the thermal insulation of ceramic-bottle construction is increased 50 percent over conventional, insulation practices.

Whatever materials are used for composite construction, a house-needy family might well begin its building project with the purchase of a 55-gallon drum of emulsified asphalt. I would not begin a construction project without this material on hand. Asphalt emulsion, of the slow-breaking, stabilizer type (bitumul), is made by more than thirty companies and is sold everywhere. It retails for 30 cents a gallon in barrel lots. As already pointed out, this versatile material will stabilize any number of filler compounds. It can also be used extensively as a wood preserver and for waterproofing.

28 PLASTICS

Plastics make up what is perhaps the best example of a composite material. Derived mainly from petroleum, coal, and wood, plastics can be molded and formed very much like ferro-cement or reinforced concrete. Low-technology-minded owner-builders may object to the consideration of plastics in this do-it-yourself presentation, but one must consider that, while the cost of traditional building materials is steadily increasing, plastics are becoming more and more plentiful. Another thing, with little practice and with little specialized equipment, one can form fiberglass reinforced plastics into an endless variety of building shapes. A composite of moldable, fluid resin over a weave of very strong glass fibers will produce a lightweight material suitable for shell construction that is virtually indestructable. Monsanto's House of the Future was built of fiber-reinforced plastic. This material was found to be strong enough to resist earthquakes, 90 m.p.h. winds, and an average of two million visitors a year during the ten years it stood at Disneyland. The double-curved shell structure, with walls $\frac{3}{10}$ inch thick, also proved strong enough to support four cantilevered wings of this cross-shaped house.

For housing purposes there are, in general, two types of widely used

248

BUILT IN AUSTIN, TEXAS BY
TAO DESIGN GROUP INC.

FLOOR PLAN

SCALE
0   10 FEET

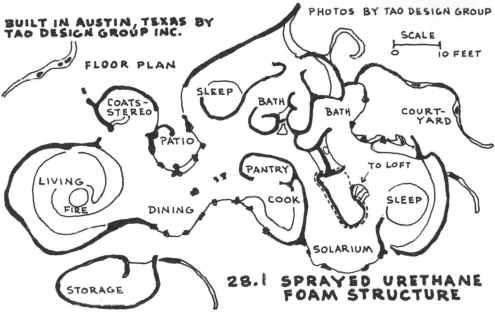

COATS-STEREO

SLEEP

BATH

BATH

COURT-YARD

PATIO

LIVING

FIRE

PANTRY

DINING

COOK

TO LOFT

SLEEP

SOLARIUM

STORAGE

28.1 SPRAYED URETHANE FOAM STRUCTURE

plastics: *thermoplastics,* which soften when heated and resolidify when cooled, and *thermosetting resins,* which harden when heated. Plastic piping (DWV, PVC, ABS) is an example of a thermoplastic; compression-molded formica is an example of a thermosetting resin.

There are, in addition, thermosetting foam plastics which are called polystyrene and polyurethane, and which are produced by the addition of an expanding agent in the chemical formulations, causing the material to form cells which expand their original volume many times.

Celluloid was the first thermoplastic developed. Since 1873, it has been continuously manufactured in a New Jersey plant, originally as a synthetic for the shortened supply of ivory billiard balls. Nearly a hundred years later, in 1966, the Architectural Research Laboratory at the University of Michigan investigated a number of building methods and building forms using foam plastic. Their report, *"Structural Potential of Foam Plastics for Housing in Underdeveloped Areas,"* failed to include a type of construction that I, at least, feel is the most suited to foam-plastic usage—the free-form house.

When foam components are mixed and sprayed on a surface, expansion of the material takes place almost immediately. The application itself requires expensive equipment and expert help. A metering device must be used to proportion the components accurately. A spray gun to apply the foam plastic must be used, adjusted to the air temperature and to the humidity at the time of application, as well as to the surface temperature of the material to be covered. No small amount of skill is required to spread a uniform coating of foam.

As Table 28.1 illustrates, foam plastic has low strength and a low modulus of elasticity but high susceptibility to heat. Foam also has low water absorption and low moisture permeability, and it has, most important, excellent properties of thermal insulation. These characteristics should influence the choice of the most suitable building form. For instance, the high susceptibility to heat certainly places a critical design limitation on foam plastic. In spite of the addition of fire retardants, foam plastic will burn vigorously. And, as the temperature rises, foam decomposes to reform the poisonous, gaseous di-isocyanates from which it was made.

The ideal plastic—strong as steel, clear as glass, and cheap as dirt—is still the stuff that dreams are made of. Meanwhile, we must

TABLE 28.1

COMPARISON OF PROPERTIES OF VARIOUS MATERIALS

	Concrete	Wood	Steel	Reinforced Polyester	Polyurethane Sprayed Foam
Density (p.c.f.)	150	35	490	82	2.5
Cost of material (cents/per pound)	2.7	6	10	48	55
Cost in place (cents per pound)	9	10	17	107	132
Compressive strength (p.s.i.)	3,000	7,200	3,300	15,000	65
Flexural strength (p.s.i.)	2,700	7,400	33,000	16,000	45
Modulus of elasticity (p.s.i.)	3×10^6	2×10^6	29×10^6	1×10^6	1,000

From the Architectural Research Laboratory, University of Michigan

work with the inherent qualities and with the limitations of the material. Because of its structural malleability, foam plastic is best used in a free-form, curvilinear design. The owner-builder can erect the armature of his free-form house, have a professional spray the wire-net form with plastic foam, and then the owner-builder can complete the shell with an interior and an exterior plaster coating.

Researchers at the University of Michigan built an interesting plastic structure of rigid foam over a folding armature. A lightweight wood lattice was first built on the ground with 2 × 2s on a 36-inch grid. Four such 27-foot-square armatures were then combined to form a dome having a 21-foot-square base. A nylon-reinforced paper was then stapled to the wood frame, and a 4-inch coating of foam sprayed onto the paper membrane.

A suggested sequence for building a free-form, earth-formed, foam-plastic core structure is illustrated in Figure 28.3. First, concrete support columns must be located and cast around the perimeter of the building. Then, a roof mound of earth is formed within the boundary of the support columns. Reinforcing bars, extending from the columns, are arched around and over the mound, and the entire above-ground surface is cast with a 2-inch layer of mesh-reinforced concrete. The concrete surface is then either sprayed with polyurethane foam, or

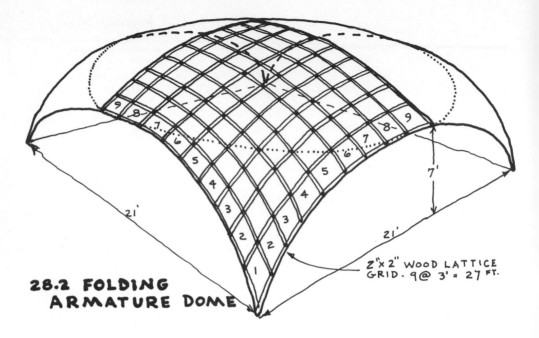

28.2 FOLDING ARMATURE DOME

2"x 2" WOOD LATTICE GRID. 9@ 3' = 27 FT.

21' 21' 7'

9 8 7 6 5 4 3 2 1

polystyrene-foam boards are laid against the concrete shell. A finish coat of cement plaster is applied directly against the plastic foam. Finally, after the foam-sandwich shell is fully cured, the earth-filled interior is removed, and interior vertical walls are plastered with waterproof cement.

In building construction, there is a wide range of use for plastics besides sprayed foam. Experimental bath units have been formed in a single moulding, including toilet, shower-tub, lavatory, floor, walls, and ceiling. Plastic window frames, gutters, piping, furniture, and wall panels are also available. Seamless polyester floors are becoming more popular in homes now that their advantages of cleanliness, chemical resistance, and ease of maintenance are becoming better known.

Plastic sealants cannot be matched for their exclusion of water and air, when several materials are being joined. Nor can silicone-based water repellents be matched for their waterproofing of above-grade masonry structures. Epoxy resins are well known for their excellent properties of adhesion. By adding 2 parts latex to 1 part cement the impact strength, the flexural and tensile strengths, the flexibility and workability, the abrasive resistance, and the adhesion quality of the cement are all improved.

Polyethylene was introduced to the building industry in 1954 by

Visking Company. In a few, short years the material has become almost universally accepted as a moisture barrier. Widths of this material up to 40 feet reduce the need for otherwise sealing floor slabs, making polyethylene especially attractive for use under concrete.

Using plastics for the exclusion of water or for the transport of water in a structure is one of the best uses of this material. Water can be excluded from a structure by using a polyethylene vapor barrier, glass-reinforced polyester wall boards, epoxy coatings, polysulfide sealants, polychloroprene glazing strips, and butyl waterproofing sheets. Water can be transported in polyethylene piping, used in polymethyl methacrylate lavatories, and disposed of in polypropylene waste pipes, which empty into polyvinylchloride sewers. *Voila!*

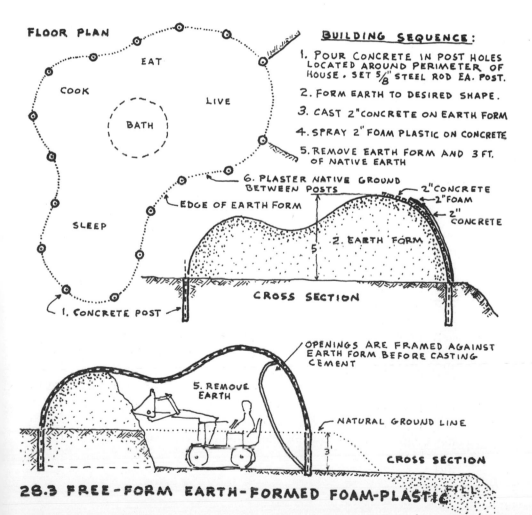

FLOOR PLAN

EAT

COOK

LIVE

BATH

SLEEP

EDGE OF EARTH FORM

1. CONCRETE POST

BUILDING SEQUENCE:

1. POUR CONCRETE IN POST HOLES LOCATED AROUND PERIMETER OF HOUSE. SET 5/8" STEEL ROD EA. POST.

2. FORM EARTH TO DESIRED SHAPE.

3. CAST 2" CONCRETE ON EARTH FORM

4. SPRAY 2" FOAM PLASTIC ON CONCRETE

5. REMOVE EARTH FORM AND 3 FT. OF NATIVE EARTH

6. PLASTER NATIVE GROUND BETWEEN POSTS

2" CONCRETE
2" FOAM
2" CONCRETE

2. EARTH FORM

5

CROSS SECTION

OPENINGS ARE FRAMED AGAINST EARTH FORM BEFORE CASTING CEMENT

5. REMOVE EARTH

NATURAL GROUND LINE

3

CROSS SECTION

28.3 FREE-FORM EARTH-FORMED FOAM-PLASTIC FILL

29 SALVAGE MATERIALS

A knowledgeable writer on building methods has stated that there will be more new homes erected during this decade than have previously been built throughout all of the history of civilization. Population increase accounts for the majority of this anticipated new construction, but no small amount of this new construction will have to be designed and built to replace the structures that have recently or currently risen only to suffer rapid deterioration and obsolescence. This is not a reference to slum clearance and urban renewal, for probably the worst domestic architecture and shabby building may be found in our suburbs—the worst architecture, certainly, in terms of its psychological detriment. Clifford Moller, in his timely book, *Architectural Environment and Our Mental Health*, speaks of architecture as ". . . an agent or a catalyst which is able to make possible a reduction of frustrations and tensions and to aid in fostering emotional stability, improved personal motivation and improved social interaction."

Obviously, the emphasis on the acquisition of physical shelter alone does not resolve the psychological stresses on people attempting to live in those shelters. Suburban-housing tracts are not planned with any degree of awareness for the psychological sensitivity of the inhabitants of those mass units. There is no more acoustical, visual, and spatial privacy and there is less genuine communication and social contact in suburban housing than there is in slum ghettos.

The practical builder-of-physical-shelter approach to human housing must, of necessity, be replaced. Yet, an architect's self-conscious, arty point of view on the subject of housing for people is not where it's at either. We must stop thinking formally. Form, in and of itself, is static and unyielding. Rather, we should think of what Moller terms as the "transactional relationship" between man and space. We should think spatially in terms of function, flexibility, and growth. It has long been known that architectural space enclosures can have a vital influence on the consciousness and on the emotional well-being of the people who live in those enclosures. Unfortunately, the achievement of a life-affirming architectural effect is the exception rather than the rule in today's suburban tract building. Most people in this solely physically-oriented housing unknowingly live lives that assure their neuroses.

The City of Detroit once paid Frank Lloyd Wright a sizable sum to diagnose that city's plight and to make suggestions for an urban renewal program. After some months of investigation and study, he made his presentation to city dignitaries at a special council meeting. "I suggest," he summarily advised them, "that you tear it all down and start over!" Perhaps this will have to be the approach taken to rejuvenating the suburbs as well as the city ghettos.

The next ten years will witness a monumental increase in new housing, and it will also witness a monumental amount of building demolition. Building salvage is already a big business. Lipsett Brothers, perhaps this country's foremost wreckers, estimates that $100 million a year is earned in salvage work, not including the additional reward from the sale of fixtures and materials such as iron pipe and electrical wire which may be gleaned from such recovery work. Buildings available for salvage are plentiful. The buildings-to-be-moved section of a newspaper lists numerous structures offered merely for their removal. Even in rural areas, buildings can be found, usually free of charge, just for the clearing of the site. State and county highway departments offer various structures on this basis, usually to clear some new road right-of-way.

There are a few salvage techniques that can be used to advantage in a wrecking operation. For safety's sake, first remove all glass windows and doors. Following the removal of all plumbing and electrical fixtures, the process of stripping takes place. This procedure removes ceiling material, paneling, trim, and molding. Properly done, minimal waste will result. One learns the process of high-grading materials in due course. The very finest salvaged material is used in the exposed sections of one's new house.

Following this stage, the high-grading operation commences from the outside and from the top down. Roof, walls, and flooring yield to removal. Structural members are set aside as is exposed outside sheathing, covered wall, roof and floor sheathing, concrete form boards, and backing material for masonry. Finally, the remains are sawed for fireplace kindling. Wood removed from old buildings is well seasoned and will not warp or shrink, but it should be protected from the weather until it is to be reused.

Members of the United Building Wreckers Local use a wrecking adz in most of their hand-salvaging operations. A ripping chisel, nail claw, hand sledge, and nail puller give additional assistance for specific aspects of the total task. I also use a modified ripping bar. A hammer head is welded to the back of the hooked end to combine hammering, nail-pulling, and prying, all in one easy-to-use tool. Tools are easily lost or misplaced in a wrecking operation, so the fewer in use at one

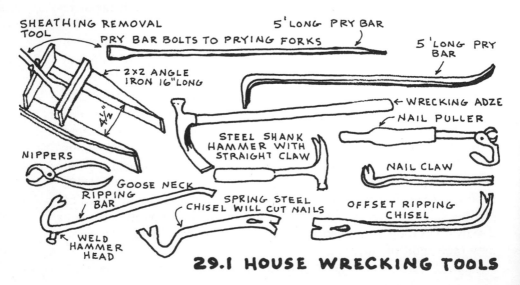

29.1 HOUSE WRECKING TOOLS

time the better. I have also improvised a sheathing-removal device, which can be bolted onto a pry bar. It consists of a pair of prying forks that lift the entire board from rafter or joist, preventing its splitting while speeding up the operation.

Scavenger personalities are a unique breed in our culture. The philosophy of salvaging is integrally linked with many of the libertarian, self-motivated, freedom-oriented concepts which form the basis of this book and its author's expression. For instance, a perceptive scavenger envisions the use of unusual material in this prospective dwelling, particularly because his vision of the structure is both unorthodox and flexible. He allows the material to contribute its own, inimitable expression to the design and the form of his building, rather than thoughtlessly discarding the material as unusable or by attempting to force the material into a preconceived mold. A dominant motive in the artistic personality is the free exercise of ingenuity and the challenge that goes with the utilization of waste materials. People put down the local scavenger as one who scabs on the system, but the reverse is more the truth of the matter. It is the scavenger who shows that the system is wasteful of its natural resources and heedless to the needs of future generations of our kind. It is not a comfortable reminder for us that our American 6 percent of the world's population annually devours 35 percent of the earth's raw materials.

Building materials are everywhere. One can find salvage, culls, dunnage, scrap, junk, and surplus items wherever one looks. One needs only to learn the fine art of scrounging to find bargains in abundance. Keep an eagle eye out for all possibilities of various materials. Haunt junkyards and auctions. Acquaint yourself with the local building-materials industries. Culls and brokens and number-two items and discontinued lines are common occurrences in any high speed factory production system. One can secure these so-called misfit materials sometimes free for their hauling. Plumbing fixtures with slight, hairline cracks are often available at half price. It took me 4,000 cull and broken adobe bricks to build the Jay's Photo Studio in Oakhurst, California. The 1970 price for the brick was 3 cents each, less than one-sixth the cost of first-quality brick.

The serious scavenger makes as his first investment an oxyacetylene torch and an arc welder. Scrap metal is relatively cheap and universally available. Used corrugated iron can be re-worked to meet a

29.2 PHOTO STUDIO   DESIGN: THE AUTHOR   PHOTO: THE JAYS

multitude of building uses. Iron pipe and 55-gallon oil drums are low in cost, easily worked, and quite versatile. Metal products are commonly available free at the local city dump. Public dumps are also a good source for broken sidewalk pavement which can be reused for walks, patio paving, and retaining walls. Telephone and electric service companies often have used poles for sale. Railroad companies sell used railroad ties cheaply. Packing crates can be salvaged with little effort. Often plywood panels are used for crating.

The use of cardboard has already been mentioned as a composite material in chapter twenty-seven. Cardboard boxes, pulverized in a hammer mill and mixed with an asphalt-clay emulsion, make a super material for floors, walls, and roofing.

The use of discarded burlap has also previously been mentioned in this text. A new twist for using burlap was introduced a few years ago by Ed Dicker of Dallas, Texas. Called the Dicker Stack-Sack method, it consists of 6- by 24-inch burlap bags filled with dry concrete mix. As the filled sacks are laid in place, two pieces of No. 3 reinforcing steel, 10 inches long, are driven through each sack. Water is then sprayed on the wall, and the wall is finished with a tack coat of $\frac{1}{2}$-inch-thick cement, sprayed on the inside and outside of the wall.

Probably the ultimate use of salvage materials should be mentioned here: Steve Baer is a geodesic dome buff who has written a booklet called *The Dome Cookbook*. In it, he tells how he removed the tops from car bodies and how he worked the resulting, metal panels into a domelike structure which he calls a zome. He first used an acetylene torch to cut out the panels, but he then found it easier to simply chop them out with an axe. One zome was erected of panels from 112 cars, including 30 station wagons and 1 van. Junkyards charged him 25 cents apiece for the car tops.

There is an interesting contrast—or, perhaps better stated, a needling conflict—in the use of salvage material like car tops with the highly sophisticated building form of the geodesic dome. The Fuller dome should be relegated to the same architectural class as the mobile home. Both are designed to be factory produced and assembled. Both have static and unyielding forms and are devoid of flexible spatial quality. The dome and the mobile home are both offered as physical shelter solutions. Indeed, they are touted as end products in building. But, as I have attempted to point out in this treatise, psychologically balanced space is not a finished end product. It is, moreover, a catalytic agent for the creation of an ever-changing visual, functional, and aesthetic order.

29.3 STACK SACK

To build with salvage material, therefore, one would anticipate that the materials themselves, in their own integral honesty, would suggest the architectural style of their use. Fuller's mathematical formulae offer only complicated systems of shelter erection that are wastefully time-consuming and functionally ineffective, while at the same time they result in little sensitivity to the psychological needs of the people who live in them. The culmination of the total turn-off is, of course, the degenerate mobile home. The scavenger-built structure would, however, evolve in free form with unencumbered flexibility. One would imaginatively anticipate its curvilinear, grottolike emergence. Space would be designed to serve many functions and to encourage change. It would respond in a flexible manner to the inimitable, unorthodox, organic vision of its scavenger builder.

30 TOOLS

Tools can be thought of as extensions of the human body. Hands (and feet) are not so much organs as they are organ holders. Human limb endings (fingers, fist, claw) can produce significant work as organ (tool) holders, and, for this reason, many people enjoy a genuine rapport with their tools. For those who do work with tools every day of their lives, this rapport is expressed in their appreciation for the clean, undesigned purity of form of the ordinary hand tool. Power equipment may do a job with exacting rapidity, but there is an irreplaceable, sensitive, human attraction to the kind of instrument that invites one's ready assessment of its heft and balance. To the tool-aware craftsman, a hardware store is a museumlike place, where one can enjoy the contemplation of quality in the make-up and workmanship of a variety of tools.

The time-honored form of classic hand tools can hardly be improved. Gadget-minded manufacturers, however, constantly try to tempt people to buy their newest, five-in-one, all-purpose tool. But attempts to change the design of the simple, top-shaped plumb bob, for instance, merely result in looking suspiciously like a toy space missile.

Choosing the right tools for one's home building can be a sometimes difficult and often confusing task. There is a seemingly inexhaustible variety of hand tools and power equipment available, but there is no guide to assist the owner-builder to make the few necessary and proper tool choices for his specific purposes. The most practical method that I have found for making this choice is to urge a prospective builder to analyze each building operation he anticipates in terms of the time required to perform the work. Then, on the basis of an evaluation of the performance required, the cost and the expected saving achieved by the use of the correct item, one will purchase the best-suited tool. Sears, Roebuck and Company's Craftsman line of tools, incidentally, can hardly be matched for price and dependability. Some specialized types of equipment may be better rented, whereas other tools which will be used later in maintenance, repair, or craft work should be purchased.

Home builders have a natural inclination to prefer one building material to another. One may be masonry-oriented, another may be wood-oriented, or a third may be salvage-oriented. Naturally, the material with which one prefers to work will influence tool selection. This chapter will discuss tools for woodworking.

As soon as the first corner stakes are driven into the ground the building process requires innumerable measurements, levelings, and lining-up activities. Tools needed for this early construction phase are a plumb bob, nylon string, chalk line, string level, 24-inch hand level, 50-foot steel measuring tape, and a 12-foot steel box tape. Be sure to buy the $\frac{3}{4}$-inch-wide box tape, not the $\frac{1}{2}$-inch size, because $\frac{1}{4}$ inch makes the difference between professional or amateur success with one's measuring tasks. A combination square should be an early acquisition. It serves a multitude of functions, and it replaces several traditional squaring tools, such as the steel framing square and the try square. This combination square can be used as a miter gauge, marking gauge, depth gauge, level and plumb, straightedge, and foot rule, as well as a try square. A sliding T-level should also be purchased. It is similar to the combination square, except that it has an adjustable blade, which can be set to any angle.

Someone once said that the hammer and saw have had more influence on the advancement of civilization than all other tools combined. Perhaps it is this impact that accounts for the many varieties of hammers and saws that one finds. Each is designed to accommodate one of a multitude of builder types who have worked with it throughout the centuries. One may purchase a 20-ounce framing hammer, a 14-ounce regular hammer, or a 10-ounce finish hammer. The claw may be curved or straight. The handle may be metal or wood. The face may be plain, bell-shaped, or checkered. Actually, only 9 hours are spent driving the 65,000 nails that secure the average wood house. The steel-handled hammer is preferred where damp-dry climate fluctuations eventually check and crack a wooden-handled hammer, loosening the head. This hammer is preferred also where exceptionally rough use is expected. The straight claw is best for ripping, and the bell-face (convex) hammer is best for general use, especially for flush nailing. Generally, the 14-ounce hammer is satisfactory for habitual use, but one should check the heft and balance prior to purchase.

The tool most used in building a wood house is, obviously, the saw. Two by fours are cut approximately 500 times to frame a house. This represents 7 days of hand sawing. The same work can be done in 30 minutes using a hand power saw. The Small Homes Council of the University of Illinois recommends the use of a small, 6-inch hand power saw in conjunction with a 10-inch, radial arm (power) saw. The usual 8-inch hand power saw is too heavy and too clumsy for the efficient accomplishment of light work, such as cutting sheathing. The true value of the hand power saw is in its lightness and in its maneuverability. About three-fourths of the cutting, shaping, and dadoing that needs to be done to build a wood house is best done with a radial arm saw.

Even with the advantage of power-saw equipment, an investment in at least five different hand saws should be made as follows: (1) a hacksaw for metal cutting, (2) a 12-inch keyhole saw for making tight, radius cuts, (3) a 26-inch, 13-point backsaw for making level, miter-box cuts, (4) a 26-inch, 5½-point rip saw for cutting parallel to wood grain (when ripping boards, keep the cutting angle at 60° and start the cut at the tip of the saw), (5) a 26-inch cross-cut saw.

An 8-point, coarse tooth saw is best for speed in sawing. The backsaw is used for work which requires smoothness. The cutting angle should be at 45°, and the cut should be started at the butt of the saw. If the wood material is green, pitchy, or wet, use a taper-ground blade. This type of blade is thinner at the tip than at the butt end and thinner on the top edge than on the tooth edge.

A 2-hour saving in wiring an average-size house for electricity can be made by investing in a quarter-inch power drill. It takes 14 seconds to hand drill a $\frac{3}{4}$-inch hole in a stud and only 6 seconds to drill the same hole using an electric drill.

Electric-drill attachments are currently popular. For the most part, these are helpful only to model-making hobbyists and should be avoided by serious home builders. True, such attachments, like the power screwdriver, reduce the time from 18 seconds to 2 seconds for driving a 1-inch, No. 10 wood screw, but a spiral-ratcheted screwdriver will place the same screw in only 4 seconds. When the time required to hook up the power drill and to change attachments is added to the actual time required to do the job, it can be seen that power tools are not always more efficient. This is especially true for short, on-the-job tasks.

There are other wood-boring operations that are best done with hand tools. The awl, for one, is the handiest and the simplest hole-making tool built. A push drill, used to drill pilot holes for screws, is handy because it can be operated with one hand while the work is held in the other.

A well-stocked tool collection should include 3 sizes of screwdrivers ($\frac{1}{4}$ inch, $\frac{3}{8}$ inch, and $\frac{1}{2}$ inch) and 3 sizes of Phillips screwdrivers (No. 2, No. 3, and No. 4). An offset, lever-handled screwdriver is worthwhile in cases where greater turning power is needed. It provides up to 10 times as much power as the ordinary screwdriver. The spiral-ratcheted (Yankee) screwdriver is a necessary investment. It spins the screw three times with one push.

Final, finishing stages in the building of a wood house require chisels for forming recesses and mortices, planes for smoothing or shaving stock, files, and rasps. Two types of wood planes should be purchased: a block plane for end-grain planing and a Jack plane, which was once called a Jackass due to its workhorselike quality.

Speed and work efficiency are promoted by arranging the tools and materials for greatest working convenience. It is wise to plan minimum tool changes. Sawing, drilling, and fastening should be done at one assembly place, and duplicate parts should be produced in one batch.

Power equipment does more than replace muscle with power. It also replaces costly skilled labor with built-in precision, thereby reducing the waste which results from inefficiency. But beware the common attitude, which many amateur builders have, that quality workmanship comes with one's acquisition of power tools. When a certain operation appears to be particularly difficult, the solution does not necessarily depend upon the procurement of new tools.

Beware, too, the five-in-one, multipurpose power tool, in which the releasing of a few wing nuts makes a lathe out of a circular saw from the drill press which, a few minutes ago, was a sander. Besides the nuisance of having to change the function of this type of power equipment (an operation entirely unsuitable to individual home building), the multipurpose aspect of this tool cheapens each separate function, making the tool next to worthless for the serious home builder.

When a builder has accumulated a collection of valuable tools, he must, next, decide how best to store them. The usual carpenter's tool box has all of the worst possible design features. Its only value is in transporting tools from one job to another. Research studies on this subject, presented by the National Association of Home Builders, found that carpenters would often spend more time hunting deep in their narrow tool boxes for a little-used tool than they actually spent working with it. Misplaced tools result in an inferior job, because improper tools are used when the proper one is missing. Ideally, a tool box should be organized in such a way that every item is clearly visible. Painted silhouettes help to get the tools returned to their proper place.

The American Plywood Association devised a compartment tool box that is far superior in design to anything else available. A modified version is illustrated below, along with a few other important tool accessories: (1) a portable workbench, which is especially useful for on-the-spot repair work, (2) a lightweight sawhorse that can be moved between stud spacings, (3) a carpenter's leather apron, which provides all the necessary pockets for four sizes of nails, a box tape, a pencil holder, a slot for a combination square, and a hammer loop. The split-leg design of this apron makes possible free leg action for climbing and kneeling.

If the owner-builder expects to continue accumulating and working with tools after the completion of his house, he may do well to plan a home workshop. It should, perhaps, be built even prior to the construction of the dwelling. The shop can be either separate or in conjunction with the main building, and can be used for storing

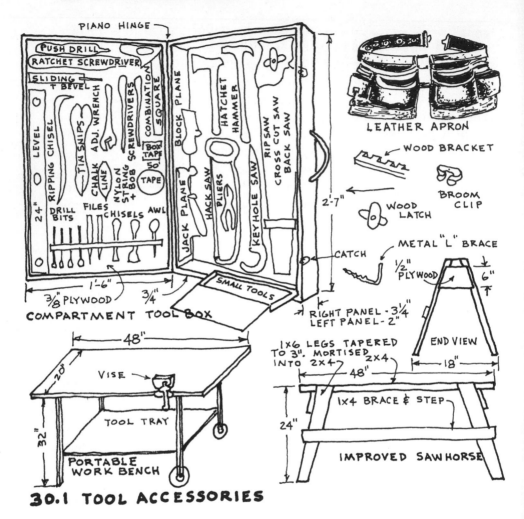

30.1 TOOL ACCESSORIES

materials, power equipment, and the large assortment of other tools which are needed to build a house. It can also be used to maintain and to repair equipment. A sturdy shop bench, wood-and-metal vise, drill press, power grinder, and storage bins for nails, screws, plumbing fittings, and so on, can offer a big morale boost to craftsman and owner-builder alike.

31 THE FOUNDATION

For the most part, it would appear that many architects today have come to expect that the home owner lives solely to manipulate modern gadgets. It is reasonably suspected that the prevailing slogan of schools of architecture may well be "design for equipment." Architects like Richard Neutra have, however, resisted this technological approach to home designing. Such practitioners caution that home planning should express the fundamental biological needs of people, rather than attempt to meet some undefinable technological needs. We should, they assert, design specifically for the edification of the human senses. They also assert, by the same line of reasoning and feeling, that the architect should become, instead, a manipulator of stimuli. It is further argued that the dramatic increase in mental disorders makes more urgent the need for housing design with a biological basis. Neutra states that each new technological invention results in urgent, new demands on the nervous system.

A home designed for equipment is an appalling prospect, but of equal repugnance is the prospect of having to pay the mortgage on an architect-designed house—assuredly an expensive though nonetheless sense-responsive structure. For many, this latter view would be just as dismal a perspective as a house designed around gadgetry. In place of either of these so-called ideals, I would prefer to have a few inconveniences and some strained senses in a house that is self-built and paid for—a house planned to fit one's site and to suit the personal requirements of one's family.

Design wise, an owner-builder is in an enviable position as the one best able to determine the family's biological space needs. Concern about the conflicting design theories of a Neutra, a Wright, or a Corbusier can be cast to the wind when one begins mixing concrete and nailing boards for one's self. Prior to construction, the owner-builder need only determine the true function of each desired or necessary building component in its relation to site conditions and climate factors. The design (or form) of the foundation, floor, wall, or roof will follow the function of these elements.

The purposes which shape the final form of the building will also suggest the materials that should be used to express those particular purposes, and these same purposes will also suggest the method of building construction. The construction of one's building foundation, for example, requires one's awareness of soil properties on the site as they relate to the weight of the building, to site drainage, to freezing conditions, and so on. When one fails to give due consideration to the functional aspects of the foundation, one may realize either extravagant waste or structural failure.

On the one hand, we naturally react with shock and sympathy to newspaper accounts reporting the premature failure of homes or public buildings which collapse, killing or injuring occupants, as a result of the use of faulty materials or of careless or even criminal construction methods. It is not uncommon for foundations built on uncompacted fill to settle, break away, or slide during heavy rain storms. Mickey Mouse plumbing or electrical installations threaten the health and safety of occupants and expose the building to early deterioration.

It is shocking, on the contrary, to also witness the forced consumption of unnecessarily large quantities of material resources to satisfy code-enforced requirements which dictate the arbitrary building systems governing contemporary housing construction. This material and the additional labor squandered to put it all into place assures one of long years of intolerable and astronomically expensive mortgage slavery. Who is to say which of these two alternatives brings the greater personal hardship—the few jerry-built homes, which eventually end up in the bottom of a ravine, or the many code-enforced, overbuilt structures, which become thirty-year millstones about the necks of unwary home buyers.

The foundation of a house can become a depository for quite unnecessary amounts of concrete. Instead of invoking fundamental formulas and common sense in the design of a foundation, this operation is usually figured haphazardly or by provincial rule-of-thumb. This elementary and vital aspect of one's building program is, additionally, subject to local, code-prescribed prerequisites. When the true function of a foundation is understood, its design, with relation to soil conditions and climate factors, will surely influence its form.

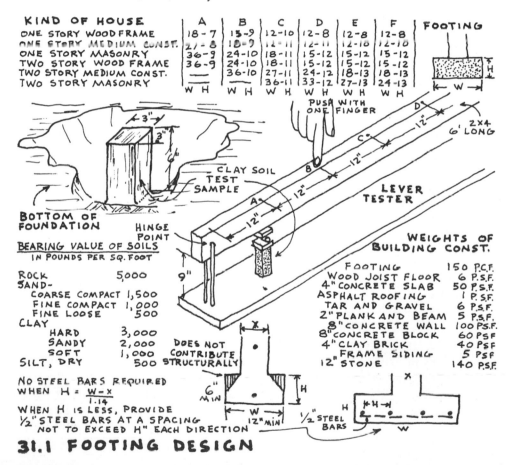

KIND OF HOUSE	A	B	C	D	E	F	FOOTING
ONE STORY WOOD FRAME	18-7	15-9	12-10	12-8	12-8	12-8	
ONE STORY MEDIUM CONST.	21-8	18-9	12-11	12-11	12-10	12-10	
ONE STORY MASONRY	36-9	24-10	18-11	15-12	15-12	15-12	
TWO STORY WOOD FRAME	36-9	24-10	18-11	15-12	15-12	15-12	
TWO STORY MEDIUM CONST.	—	36-10	27-11	24-12	18-13	18-13	
TWO STORY MASONRY	—	—	36-11	33-12	27-13	24-13	
	W H	W H	W H	W H	W H	W H	

BOTTOM OF FOUNDATION

CLAY SOIL TEST SAMPLE

PUSH WITH ONE FINGER

LEVER TESTER

HINGE POINT

BEARING VALUE OF SOILS
IN POUNDS PER SQ. FOOT

ROCK	5,000
SAND-	
COARSE COMPACT	1,500
FINE COMPACT	1,000
FINE LOOSE	500
CLAY	
HARD	3,000
SANDY	2,000
SOFT	1,000
SILT, DRY	500

DOES NOT CONTRIBUTE STRUCTURALLY

WEIGHTS OF BUILDING CONST.

FOOTING	150 P.C.F.
WOOD JOIST FLOOR	6 P.S.F.
4" CONCRETE SLAB	50 P.S.F.
ASPHALT ROOFING	1 P.SF.
TAR AND GRAVEL	6 P.S.F.
2" PLANK AND BEAM	5 P.S.F.
8" CONCRETE WALL	100 P.S.F.
8" CONCRETE BLOCK	60 PSF
4" CLAY BRICK	40 PSF
FRAME SIDING	5 PSF
12" STONE	140 P.S.F.

NO STEEL BARS REQUIRED WHEN $H = \dfrac{W-x}{1.14}$

WHEN H IS LESS, PROVIDE $\frac{1}{2}$" STEEL BARS AT A SPACING NOT TO EXCEED H" EACH DIRECTION

6" MIN

12" MIN

$\frac{1}{2}$" STEEL BARS

31.1 FOOTING DESIGN

A discussion here of foundation consists of three parts: (1) the bed, which is the earth support, (2) the footing, or the widened part of the construction, which rests on the bed, and (3) the wall, or the structural part, which lies upon the footing. An owner-builder can easily design the foundations of his house by knowing the load-carrying capacity of the bed (soil) support and by knowing the approximate weight of the building and its contents. The table in Figure 31.1 indicates what are the load-carrying capacities of common soils. One can also perform a simple test to evaluate the strength of soil with critical, high-clay content. A small block of soil is taken from the area of earth which is to be the support of the foundation (see drawing). The bottom and the top of the test sample must be trimmed squarely and must be set in a simple lever tester, as illustrated. If the soil block crushes when pushed at point A with one's index finger (about 20 pounds of force), then the foundation soil must be considered too soft to support a house. Move the finger along the lever arm until the soil sample crushes. The point of failure can be correlated with the requirements for the footing width and depth, as indicated in the table. The table serves to illustrate the relationship between the weight of the house and the thickness of the foundation wall and the soil sample. On firm, well supported foundation beds, the footing should be twice wider than the foundation wall, and it should be as deep as the wall is wide.

Factors of frost line and ground water are other determinants that control the choice of materials and the design of one's foundation. The frost line through the northern United States varies from 4 to 6 feet in depth, and, in most areas of the central states, it averages from 2 to 4 feet. Foundation walls must extend below these depths if freezing is to be avoided. The alternate expansion and contraction of the earth during freezing and thawing activity may heave the footing, damaging the foundation wall and possibly the superstructure itself. An alternative solution is to use rock ballast, in a manner similar to its use in railroad beds, under the foundation and under the slab flooring to insure maximum drainage. Gravel-filled trenches under the footing have also been successfully used for the drainage of ground water. A gravel bed will support 6 tons of building weight per square foot.

Where basements are to be included in the building, all manner of complex foundation problems arise. High water table, moisture from

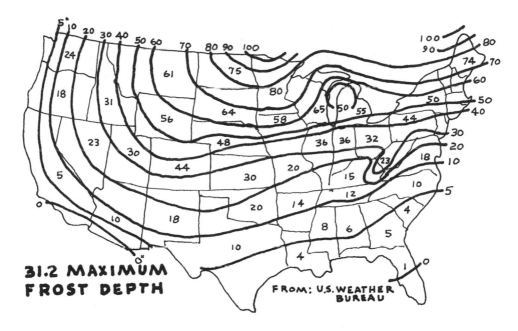

31.2 MAXIMUM
FROST DEPTH

FROM: U.S. WEATHER
BUREAU

rain and snow, or even underground springs all contribute to the general nuisance value of basements. Basically, the problem is that water seeks its own level, and one must either drain the water off, which is seldom possible on the average basement site, or one will inadvertently build what amounts to a swimming pool in reverse.

The additional cost of a basement in a one story house of average size is $2,000 or more, depending upon the cost of the excavation. At least 10 percent more living space above grade can be had for the same money. Basementless houses appeal to those people who are low-cost oriented because of the smaller initial investment required of them. Older people, for whom stair climbing demands extra effort and poses physical hazard, react favorably to homes without basements.

Soil conditions, frost-line depth, and ground water content are the main factors which influence footing design, but, in addition, the contour of the site and the distribution of the building weight must also be considered. To assist the owner-builder in his choice of a foundation system, various types of footings are outlined below.

My favorite method of building foundations is, at the same time,

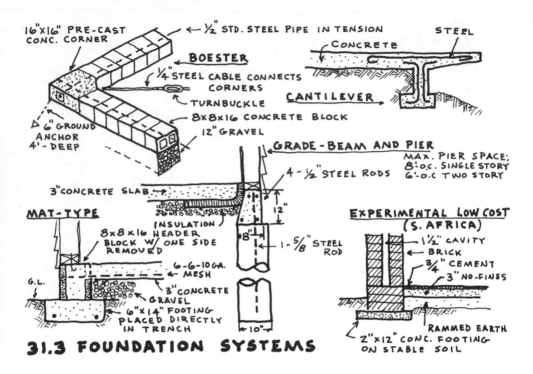

31.3 FOUNDATION SYSTEMS

the simplest and the least costly. A footing is poured directly into a trench. Consistent footing height can be ascertained by leveling a series of wooden grade stakes driven directly into the foundation bed. After this footing is poured and partially set up, a single row of header-type concrete blocks is set around the building perimeter on this poured footing.

With a mat-type footing, where the slab floor is reinforced by wire mesh to act as a unit distributing building loads over the entire surface, the inside wall of the concrete header blocks is broken out to receive a continuous pour of concrete. Thus, the blocks double as a screed runner and as a form for a poured floor, and the mat acts as both foundation and footing. Mat footings are generally used in areas where there is little or no freezing. In other areas, mat footings can be designed to bear directly on concrete piers sunken below the frost line.

The use of concrete piers accompanies grade-beam foundation construction which is a fairly new method and is known to have many economic advantages over the traditional, continuous-footing foundation construction. According to foundation-cost comparisons made at Pennsylvania State University, a continuous-footing foundation, using concrete block or poured concrete walls, will cost about 25 percent more than a slab-on-ground floor built with a grade-beam and pier foundation. A saving in cost by use of grade-beam construction is especially apparent in homes built in northern climates.

The first step to build a grade-beam foundation is digging pier holes 6 to 8 feet apart along the perimeter of the house. The holes should be dug below the frost line, with the bottoms of the holes well established in good, load-bearing soil. Concrete piers are made accordingly: they are 10 to 12 inches in diameter with a $\frac{5}{8}$-inch reinforcing rod protruding 11 inches from the top of the pier into the space, which receives the poured concrete grade beams. The strength of the grade beam depends upon well placed reinforcing rods. Two $\frac{1}{2}$-inch rods should be placed horizontally at the top and the bottom of the beam. Though poured on the ground between wooden forms, a grade beam is actually supported by the piers.

A stone-filled foundation trench, under a continuous footing and extending below the frost line, was suggested by Frank Lloyd Wright in his book *The Natural House.* Actually, this method was used a century ago by masonry-wall builders in the northeast. Stone ballast, similar to that used on railroad beds, has been used successfully for masonry-wall support in areas experiencing deep frost penetration. It will support exceptional weight loads and at the same time will provide drainage, preventing frost damage.

Another variation of the stone ballast idea for foundation support comes from Carl Boester of Lafayette, Indiana. Boester developed a mortarless, concrete-block foundation system supported on gravel fill. Steel rods and metal corner anchors permit an entirely dry block assembly to be built without mortar. In areas of the country not supplied with electricity, this method is a good option. Half-inch steel pipes are simply placed through the center core of standard concrete blocks and tightened at each corner of the building; that is, they are post-tensioned, and, in addition, $\frac{1}{4}$-inch steel cable is made diagonally taut at the building's outside corners.

Possibly the most significant advance of the century in foundation design has come out of a successful research collaboration between

The National Forest Products Association, The American Wood-Preservers Institute, and The U.S. Forest Service. In this instance, stone-ballast-foundation construction is employed to its maximum utility. A two-story, brick-faced structure with a full basement was built to bear on a single 2- by -8-inch wood footing, resting on a 6-inch-thick pad of gravel. No concrete whatsoever was used. Even the basement walls consist of plywood-faced, stud-wall panels. Of course, all wood members used below natural grade must be pressure treated to meet the requirements of The American Wood-Preservers Institute. Look for the A.W.P.I. quality mark before purchasing the wood.

The key to designing an all-wood foundation system is in one's provision for adequate drainage. When building a full basement, a 4-inch-thick gravel fill and a sump must be installed below the floor

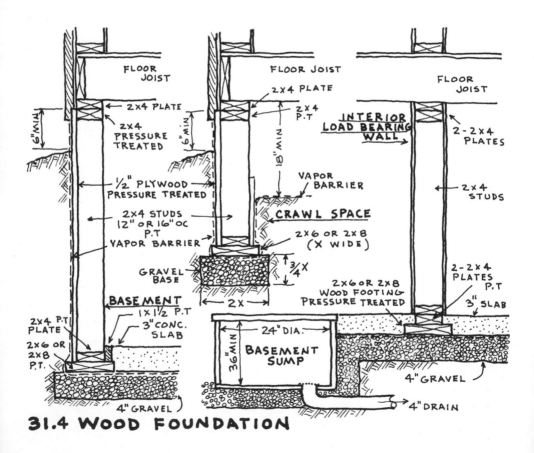

31.4 WOOD FOUNDATION

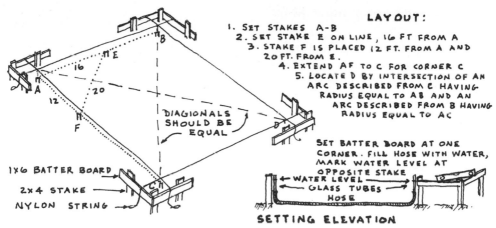

LAYOUT:

1. SET STAKES A-B
2. SET STAKE E ON LINE, 16 FT FROM A
3. STAKE F IS PLACED 12 FT. FROM A AND 20 FT. FROM E.
4. EXTEND AF TO C FOR CORNER C
5. LOCATE D BY INTERSECTION OF AN ARC DESCRIBED FROM C HAVING RADIUS EQUAL TO AB AND AN ARC DESCRIBED FROM B HAVING RADIUS EQUAL TO AC

SET BATTER BOARD AT ONE CORNER. FILL HOSE WITH WATER, MARK WATER LEVEL AT OPPOSITE STAKE

WATER LEVEL
GLASS TUBES
HOSE

DIAGONALS SHOULD BE EQUAL

1×6 BATTER BOARD
2×4 STAKE
NYLON STRING

SETTING ELEVATION

31.5 FOUNDATION LAYOUT

slab, allowing all water to drain from under the slab. This drainage system provides a positive method of eliminating the hydrostatic pressure which is the primary cause of wet basements.

All structural members in an all-wood foundation function to maximum capacity. When the bearing capacity of the gravel base and the supporting soil are 3,000 pounds per square foot respectively, 2- by -6-inch footings are adequate to support the superstructure. Bearing capacity for 2 by 8s would be 2,000 pounds. Vertical, 2- by -4-inch or 2- by -6-inch framing members are sufficient to support live and dead loads as well as to contain the lateral forces of earth back-fill, which tend to push in on the wall. One-half-inch plywood wall sheathing also resists this outward soil pressure occurring at the lower sections of the wood wall. The 3-inch concrete base (slab) adequately opposes lateral hydrostatic force, which has a tendency to thrust walls inward.

The Federal Housing Administration has recently approved all-wood foundations for mortgage loans, and the Department of Housing and Urban Development includes this system in their evaluation of low-cost building proposals for Operation Breakthrough. Closely watched, time-motion studies observed that, with a five-man crew, a wooden-basement foundation could be erected in $1\frac{1}{2}$ hours as compared with the $10\frac{1}{2}$ hours required for the construction of a regular, concrete-block basement. This saving in time represents a materials saving of $280 and a saving of 60 man hours of labor.

Whichever system of foundation construction an owner-builder decides to use, some simple construction practices can be advantageously employed to eliminate error and unnecessary labor.

By erecting batter boards, building corners, building perimeters, and floor levels can be preserved during excavation. Floor levels are established and marked on the batter boards. Equal elevation is indicated by the level in a garden hose filled with water. The application of the square-of-the-hypotenuse theorem provides a simple test for exactly determining right-angle corners for the foundation. Mark 16 feet along the cord on one side of the batter-board stakes, 12 feet along the other cord, and the line connecting the two markers is the hypotenuse of a right triangle and should be 20 feet. Curvilinear buildings can be laid out from a single radius or from a series of radii. For an example, see the spiral wall slip form, Figure 11.6.

At times, special ingenuity is required to solve foundation problems on unusual sites. For instance, in areas where only muck or silt exist as bearing soil, a raft footing becomes necessary. In this case the entire building is supported on a concrete raft, which floats on the wet soil. The building is designed so that its load equals the muck it displaces.

Hillside sites always require that special attention be given to the foundation system. Concrete retaining walls may be needed. A cantilever foundation system is often used to advantage on hillsides. Wright called the cantilever "the most romantic, most free of all principles of construction," and he used the cantilever in most of his famous buildings, including The Imperial Hotel in Tokyo and The Johnson's Wax Company Tower. This building emulates tree form. In structural terms, a tree is a beam vertically cantilevered out of the ground. It maintains stability against wind pressure and snow loads by means of the restraint of earth against its roots. There is a subtle balance between the bearing force of the earth and the tree (beam). With Wright's tower, a central, vertical core and a cantilevered foundation system function as the only structural support.

It is the problem site that most challenges the concept that form and function are one. In any building circumstance, where this concept is operative, the foundation must necessarily act as a part of the downward thrust of roof and wall and not as a separate and independent pad upon which the house is set. The concept of a building having a tap root (foundation) growing outward from its site was not originally a simple idea. It required a master builder like Wright to imaginatively capture and tame this organic, architectural theme.

32 FLOORS

> He who considers things in their first growth and origin . . . will obtain the clearest view of them.
>
> Aristotle, *Politics*

There is some practical value in obtaining a clear view of house evolution. Every building component, whether floor, wall, or roof, has a unique development from prehistoric to modern times. We can readily trace this development and often discover the moment in human history when man discarded a functional and economical solution in favor of some embellished, complicated, malfunctioning form. It may be wise to return, with a few technical improvements, to the functional solutions.

A floor, for example, should not be considered as being merely a hard, shiny, easy-to-clean, operational surface for the efficient housewife. Nor is it to be considered just a structural diaphragm bearing 40 pounds of weight per square foot. It should more properly be thought of as an interior field, for, anthropologically speaking, the floor belongs to the farmer phase of our development, preceded only by the hunter's roof tent and the herder's wall fence. Our farming predecessors built floors level like a field—a dry surface for storing crops. As a fitting adjunct to his floor development, the early farmer elaborated on chair, bed, and couch forms. A three-legged stool counterbalances the uneven ground, whereas a level floor requires four-legged furniture. The farmer's long and happy reign in which he provided

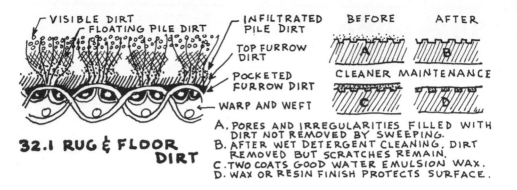

32.1 RUG & FLOOR DIRT

VISIBLE DIRT
FLOATING PILE DIRT
INFILTRATED PILE DIRT
TOP FURROW DIRT
POCKETED FURROW DIRT
WARP AND WEFT

BEFORE   AFTER

CLEANER MAINTENANCE

A. PORES AND IRREGULARITIES FILLED WITH DIRT NOT REMOVED BY SWEEPING.
B. AFTER WET DETERGENT CLEANING, DIRT REMOVED BUT SCRATCHES REMAIN.
C. TWO COATS GOOD WATER EMULSION WAX.
D. WAX OR RESIN FINISH PROTECTS SURFACE.

his own simple accouterments for himself was eventually replaced by the ascendancy of the exploiting artisan who added to the shelter embellished furnishings, wall openings, and drain lines.

No matter which element of a house one chooses to study, there is always a time in history when the pure-and-simple and the functional aspects become complicated and costly. This happened to floors when Sidney Lanier's legendary prince, enamored of the feel of leather, demanded that it be spread on every floor upon which he walked. We are doing likewise today for suburbanites who demand wall-to-wall carpeting and polished floor tile. It is surmised, facetiously, that leather shoes were developed to economize on leather flooring. Perhaps, as people accustom themselves to proper footwear, the current fad of wall-to-wall carpeting will pass away. Costs and aesthetics aside, if people could view a magnified section of carpet, their hygenic orientation would compel them to rip out the carpet and to walk on the wood floor with carpet slippers instead. One need only realize that from 85 percent to 95 percent of all the dirt that accumulates in a room is in the carpeting. Owing to the structural formation of warp and weft, the rows of carpet tufts have an enormous capacity for storing dirt.

The author's favorite architectural critic, Bernard Rudofsky, sums up his feelings about floors:

Floors are meant to be walked on. But wearing shoes at home has not only made us indifferent toward dirt, it has blunted our sense of touch. The admonition to take off our shoes before entering a house (a Japanese house, for example) is felt, if ever so vaguely, as humiliation—as if one were told to adjust one's fly. . . . Without that provision for an extra kick, the heel, we feel incomplete. The crushing heel of the male, the vicious spike of the female are part of their armor against life. We have progressed so rapidly in recent years that we are now catching up with ancient technology.

Any building researcher worth his salt will first investigate the sheltering forms of ancient technology. This investigation should be made before assuming that some totally new floor material needs to be fabricated.

The floor is one of the most elementary of man's shelter components. After the hunter and the herder developed roof and wall protection, the farmer began to concern himself with the moisture and cold that rise from the ground. In most parts of the world, this problem is still essentially unsolved. Instead of finding a satisfactory floor covering which would permit him to live on his native ground, man has invented numerous devices which have kept him elevated above the ground. The cold and moist floors of early history are responsible, it seems, for the invention of chairs, tables, and beds.

It is interesting to see how well the Japanese have done without over-furnished interiors, thereby successfully solving the floor problem. The floor in a Japanese house is invariably made of wood raised several feet off the ground, and surfaced with a 2-inch-thick insulation mat. Consequently, the Japanese people are able to eat, sleep, or sit directly on this relatively warm, matted floor. Their houses are simple, though highly refined. Their lives are, therefore, less burdened, without concern for the acquisition, care, and clutter of a houseful of furnishings. But, more important, their health and posture are not impaired by raised sitting surfaces, which are unquestioningly accepted by western society.

Staff members at the John B. Pierce Research Foundation made it a point to study all of the primitive and traditional methods of flooring before issuing a 5-year research report on floors. One researcher sums up a return to the solid primitive:

> Nature has endowed us with a wealth of new materials which man, through his ingeniousness, has learned to process in highly complex ways so as to better serve his needs for shelter. He has

not been content with such crude productions as the thatched roof or the pueblo of adobe clay but has transmuted Nature's products into more efficient devices. The cost added in manufacture in these processes has steadily mounted to the point where it may be well to hesitate and turn our efforts more toward simplifying and reducing the cost of this conversion of basic raw materials to finished products. There may be good reason to believe that, so applied, the science of soil mechanics—the same science that has been able to stabilize earth around oil wells by running an electric current through it, that has stabilized earth embankments by injection of current, that has built better levees and airport runways through the use of earth-bituminous mixtures—can ultimately produce a satisfactory house floor out of earth.

Although stabilized earth floors have been used extensively in the past, especially in South Africa and India, the material does not provide a cleanable, highly wear-resistant surface. Correct proportioning of a stabilizing agent, like bitumul or cement, is critical when earth is used as a base material.

For low-cost applications involving light wear, sawdust-cement floors have been proven suitable, giving greater warmth and quietness than conventional concrete floors. Sawdust from softwoods should not contain quantities of resins, for resins prevent the set of concrete. The addition of lime to the sawdust-cement mixture improves workability. Sand is frequently added to reduce high shrinkage. A proportion recommended by the South African Building Research Institute consists of 1 part cement, $\frac{1}{4}$ part lime, $1\frac{1}{2}$ parts sand, and 2 parts sawdust. The mixture should have the consistency of moist soil, sufficiently dry so that the trowel does not bring the cement to the surface as a skin. The mixture should be damp cured for 10 days and be thoroughly dried before applying linseed oil or wax polish.

Functionally speaking, it would be well nigh impossible to find a more suitable additive for concrete floors than common, inexpensive, emulsified asphalt. Where asphalt emulsion is used in place of water for the gauging liquid, the cement-hardening action is slowed down, making the floor less brittle. A floor of this type has a wear-resistance comparable to hard maple. It is warm to the touch, with 10 percent less thermal conductivity than concrete.

Until the John B. Pierce Foundation published formulas for what it called its comfort-cushion floor, the only published material available came from England. The English applied their asphalt-concrete mixture on a subfloor of concrete or wood. A typical mixture consists of 80 pounds of cement, 11 gallons of emulsion, and 310 pounds of small aggregate, $\frac{1}{8}$ to $\frac{1}{2}$ inch in size. After a 24-hour damp cure, the floor surface can be waxed and polished. The Pierce Foundation comfort-cushion floor formula consists of 34 parts stone, 22 parts sand, $\frac{1}{2}$ bag of cement, and 1 gallon asphalt emulsion. It figures that 7 gallons of asphalt emulsion and $3\frac{1}{2}$ bags of cement are used in each cubic yard of concrete. This mixture is dumped and spread on the work area with a rake or shovel. A 125-pound roller is then used in two directions until the surface becomes smooth. Meanwhile, a batch of the same mix, excluding rock, is prepared. This is then spread over the surface and rolled slightly to float finish the material.

The asphalt additive in comfort cushion floors is said to provide both resilience and moisture protection. Resilience is measured by the immediate yielding of the floor surface to the impact of the foot and to the surface's return to its original position. On the basis of tests made at the National Bureau of Standards, resilience values may be questioned, but there is certainly no doubt that asphalt provides an adequate moisture barrier in a slab floor. This is important. Vapor migration is, perhaps, one of the main objections to concrete floors. This is a phenomenon only recently understood by engineers, as water vapor is an elusive, invisible gas. Vapor travels from one area to another whenever a difference of pressure exists between the two. Vapor pressure increases with both temperature and humidity rise. Radiant-heat pipes installed in concrete have the dismal effect of increasing moisture transmission through the slab, when no vapor barrier is installed. A highly impermeable membrane, such as a 55-pound roll of roofing or polyethylene plastic, should be placed under the concrete slab to prevent vapor migration.

Capillarity is a second problem encountered in floor-slab construction. The amount of moisture that can be transmitted from the ground by capillary, wick action is usually underestimated. Soil experts claim that the relative humidity directly beneath the slab stands at 100 percent. In rooms of low humidity, moisture will tend to migrate from the high to the low; that is, up through the floor. Soil-moisture content varies in direct relation to the fineness of the soil. Several inches of coarse, washed gravel will adequately interrupt capillarity. Gravel beneath the slab also serves as a thermal break between slab and ground. Since heat absorbed from the room is not dissipated as rapidly to the ground, winter comfort is increased. Heat is transmitted about four times as fast through damp silt or clay soil as through dry gravel.

Interesting variations of the gravel-filled slab base have been made by builders in Southern Rhodesia and South Africa. They use the no-fines concrete mixture mentioned in chapter twenty-two. A 3-inch layer of 8 parts crushed stone to 1 part cement is spread over a tamped earth filling. This no-fines surface bed is immediately covered with a $\frac{3}{4}$-inch cement screed, merely for a smooth finish. Thermal-insulation properties are said to be better with this mix than that made with dense concrete, and capillarity is reduced. There is cost economy, moreover, in this method, since less cement, as well as no sand, is used in the mixture.

Ideally, the best guarantee against capillarity is a continuous air space between floor and ground. A low-cost floor of this nature was developed a few years ago by Dr. Billig of the Central Building Research Institute in India. Although described as a light-duty floor, it was subjected to loading stresses of up to 450 pounds per square foot, without showing any sign of distress. (Most building codes in America require a 30 to 50 pound per square foot minimum. In conservative, residential occupancy the furniture loads seldom exceed 15 pounds per square foot, uniformly distributed.) The C.B.R.I. floor

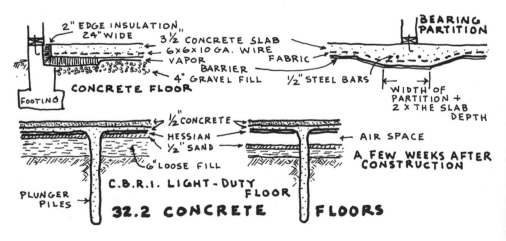

32.2 CONCRETE FLOORS

consists of a 1-inch-thick, lightly reinforced concrete slab resting on plunger piles. To make a hole for a pile, a crowbar is driven into the ground to a depth of 3 feet. The hole thus made is filled with fine concrete. The piles are spaced on 3-foot centers. The slab consists of two layers of concrete, each $\frac{1}{4}$ inch thick, spread over hessian, a form of burlap. After a few weeks, the loose earth filling under the slab settles, and an air space is formed under the slab, which finally rests on the concrete piles. Effective heat insulation results, with the floor being cool in the summer and warm in the winter. The dead weight of the floor is on the order of one-fourth that of a conventional, concrete floor. Where a no-draft, under-floor air-circulation pattern is planned, as illustrated in the hypothetical house design, Figure 2.1, an owner-builder could do no better than to install a plunger-pile floor. An English variation of this flooring system specifies the use of a rototiller to loosen the earth before pouring concrete. Supposedly, soil stirred in this manner compacts more after the slab has settled and will provide a larger air-space cavity.

Given a well-drained building site, an elevated, air-spaced, or gravel-filled subfloor, and a continuous, impermeable moisture barrier, there is no functional reason why a concrete floor cannot be adequately resilient, warm, and dry.

The author has found that good quality concrete, relatively impermeable to moisture migration, can be produced even without a moisture barrier. First, the mixture proportion must be correct. Shrinking and cracking are caused almost entirely by too much water in the mixture. With a 5-sack mix (5 cubic feet of cement for each cubic yard of concrete), no more than 6 gallons of water should be used. A good proportion is 1 part cement, $2\frac{1}{4}$ parts sand, and 3 parts rock.

Immediately after the slab is poured, the concrete should be thoroughly consolidated by vibration, spading, or tamping. It can then be screeded to proper grade. The screeding operation is immediately followed by floating or darbying, which is done with a specially constructed, long wooden trowel. The long darby is used to embed coarse aggregate in preparation for hand floating and troweling. It helps to level the surface and provides further compaction. Final troweling is done only after the concrete has hardened sufficiently so that moisture and fine particles do not rise to the surface. A wooden trowel is first used, followed by a steel trowel when the concrete is so hard that no mortar accumulates as the trowel draws over the surface. Concrete floors should be moisture-cured for at least 3 days. Cement tools needed for floor finishing by the owner-builder are illustrated below.

An improperly laid, concrete floor can be unhealthful and discomforting to occupants. Improperly laid concrete floors have been so common that most home buyers refuse to have them installed. The usual complaint against this flooring is that the lack of resiliency in concrete causes leg and back pain. Yet, according to tests at the National Bureau of Standards, there is no more give under heel or foot on wood or asphalt tile than there is on concrete. Human fatigue is affected more by the difference in resilience between rubber- or leather-heeled or soled shoes than it is between a wood or concrete floor.

The owner-builder who is completely sold on wood floors will find many economical and structural advantages to having the wood surface installed over a thin concrete subfloor. Besides its attractive appear-

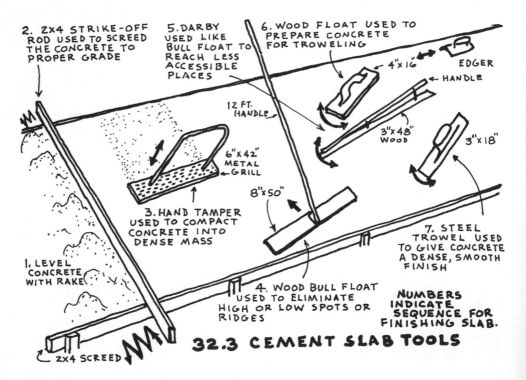

2. 2X4 STRIKE-OFF ROD USED TO SCREED THE CONCRETE TO PROPER GRADE

5. DARBY USED LIKE BULL FLOAT TO REACH LESS ACCESSIBLE PLACES

6. WOOD FLOAT USED TO PREPARE CONCRETE FOR TROWELING

4"x16"

EDGER

HANDLE

12 FT. HANDLE

3"x48" WOOD

3"x18"

6"x42" METAL GRILL

3. HAND TAMPER USED TO COMPACT CONCRETE INTO DENSE MASS

8"x50"

7. STEEL TROWEL USED TO GIVE CONCRETE A DENSE, SMOOTH FINISH

1. LEVEL CONCRETE WITH RAKE

4. WOOD BULL FLOAT USED TO ELIMINATE HIGH OR LOW SPOTS OR RIDGES

NUMBERS INDICATE SEQUENCE FOR FINISHING SLAB.

2X4 SCREED

32.3 CEMENT SLAB TOOLS

ance, hard wood offers excellent wearing qualities for minimum up-keep. Wood also has low heat conductivity and is readily adaptable to owner-builder construction. However, wood has the unfortunate tendency to shrink and to swell as its moisture content varies, so special precautions must be taken when it is used over a concrete subfloor.

Hot melt mastics with asphalt base have been widely used to bond wooden parquet flooring to concrete. The tar is merely poured hot onto the concrete subfloor (a gallon covers 30 square feet), and this hot tar is then leveled with a trowel tooth before setting the parquet blocks. Cut-back mastics have also been used with success in this operation. A mixture of emulsified asphalt, sand, and cement must first be applied to level irregular concrete surfaces. Recent developments in concrete-wood flooring suggest that the builder embed redwood or pressure-treated 2- by -2-inch sleepers (screeds) on a mastic-covered concrete subfloor. These wood strips are spaced 16 inches apart, with staggered lengths of from 18 to 48 inches.

Economically, wood can compete with concrete for flooring only under unusual circumstances—as on a hillside site and only when some of the newly developed framing and covering methods are employed. The obvious way to save on wood flooring is by the correct choice of lumber grade and size. The allowable span of a floor joist, for instance, depends upon either stiffness or strength. Lower grades of lumber can often be used where the strength factor is higher than necessary, but where stiffness requirements are met. In Canada, a table entitled "Housing Standards" has been issued which indicates a number of different grades of wood in relation to their strength and stiff-ness. From this table, one can select the most economical material for spanning any required distance. These Canadian-housing standards also require less design loading for bedrooms than for other parts of the house. With bedrooms representing about 40 percent of the total floor area, the saving between 24-inch joist spacing and 16-inch joist spacing becomes significant.

In Blacksburg, Virginia, in 1967, Homer Hurst built an experimental house containing a number of innovative flooring concepts. He achieved an integral T-beam action by glue-nailing the hardwood floor finish directly to the subfloor. It was found that hardwood flooring laid parallel to floor joists on top of subflooring is at least 50 percent stiffer than the joist alone. Joists $1\frac{1}{4}$ inches by 8 inches were lap-jointed with hardened screw nails to span 40 feet at three equal support joists.

The Douglas Fir Plywood Association has taken the lead in developing some really functional wood-flooring systems, like stressed-skin plywood panels which span 8 feet between bearing members. The top layer consists of $3/4$-inch plywood, and the bottom layer can be built of $1/4$-inch plywood. Panels are made up in a jig and have only to be set between beams.

A single-thickness plywood has also been developed by D.F.P.A. It is called 2.4.1 and consists of seven plies of wood totaling $1\frac{1}{8}$ inches thick. This extra-thick plywood will span 4 feet, making for a simple, few-piece girder and a simple blocking network.

There are over thirty choices of flooring materials, based upon the various classes of flooring, their thicknesses, and type. The old-fashioned standby of linoleum for the kitchen, ceramic tile for the bath, and wood everywhere else has been replaced by some complicating alternatives.

TABLE 32.1

FLOOR COVERINGS

P = poor F = fair G = good E = excellent	Wood	Concrete	Ceramic Tile	Linoleum	Cork	Rubber	Asphalt	Vinyl Asbestos	Vinyl Regular	Vinyl Homogenized
Installation cost sq. ft.	.20	.15	1.00	.50	.60	.80	.30	.50	.40	1.00
Use below grade	no	yes	yes	no	no	yes	yes	yes	no	yes
Thermal conductivity Btu/Hr/Sq. Ft./°F/in.	.80	11.0	5.0	1.5	.50	5.3	3.1	3.1	1.4	5.3
Underfoot comfort	G	E	E	G	E	E	F	G	G	E
Quietness	F	P	P	G	E	E	F	G	G	E
Ease of maintenance	F	P	E	E	F	G	F	G	E	E
Grease resistant	G	P	E	E	F	G	P	E	E	E
Relative max. static	E	E	E	G	G	E	F	F	G	E
Load. lbs./sq. in.	—	—	—	75	75	200	25	25	75	200

In choosing a surface covering for concrete floors, it is important that the material be resistant to moisture and alkali. Even with the use of the best vapor barrier available, some moisture is bound to penetrate the concrete floor, dissolving the alkaline salts in the concrete. Alkali attacks linseed and other vegetable oils, so many tile adhesives and linoleums become brittle and discolored when laid on concrete.

A secure anchoring of all wood components is the most important consideration in achieving a solid-wood floor. Threaded sinker nails have proven best for fastening underlayment. This one practice prevents springiness, uplift, horizontal shifting, warpage, and nail-head popping. Differential shrinkage of joists and girders can be prevented by the use of solid block bridging or by diagonal struts.

Table 32.1 summarizes volumes of literature published on the subject of floor coverings. In the final analysis, the important thing that the owner-builder should realize in his choice of any type of flooring material or in his choice of a floor-building system is that his selection must be made and executed on the basis of its functional performance rather than on the usual pattern of selecting flooring and floor-building systems by volume specifications.

33 WALLS

Man's first wall construction was a stockade around his village. Its function was to protect village occupants from intruders and to contain village livestock. In the evolution of house forms, wall construction by herdsmen succeeded the roof-tent constructions of hunting ancestors. Originally, the herder built his wall by driving posts into the ground and weaving wattle between. Later, he pressed stiff mud into the wattle, a technique which was the direct predecessor of our current and perennially popular half-timbered wall style. Finally, a few thousand years before Christ, the wall builders invented the brick, an event which essentially brings us to the modern scene.

As contemporary man continues to place brick on brick, symbolically reforming the stockade wall around his living space, he armors himself against life's ravages in much the same manner that his pastoral ancestors had to for survival. The idea that walls function for protection has persisted until the present, even though the early significance of these devices has been lost to our present-day, nonpastoral communities. Some self-conscious design reformers have, however, started a campaign for bringing the outdoors in. Result: the picture window has become a stock item in housing, and indoor planters have been substituted for the outside feeling, just as the concrete patio has been substituted for the inside feeling.

288

It's about time we reexamine the wall in terms of its purpose and function. A house can be thought of as having many purposes, but primarily it is a contrivance for regulating a segment of our environment. This controlled segment of the environment is enclosed with roof, floor, and walls so that weather factors, such as air movement, humidity, precipitation, temperature, and light, may be regulated. Walls are structural membranes separating the indoor from the outdoor environment. On the other hand, the greater the difference between the outdoor and indoor environment, the more elaborate must be the wall's inherent properties to cope with this differential. The more important characteristics that a wall must have include strength and durability, a control of the flow of heat, moisture, air, and good design at low cost. These are the more significant properties of wall construction that will concern us in this chapter.

Generally speaking, about one-half of the cost of a house is the cost of its materials; the remainder is labor costs. With this fact in mind, one would naturally assume that builders have a thorough knowledge of the nature and properties of all possible building materials. The functional performance of a wall material is the number one consideration for the wall-building phase of construction, but, at the same time, the most economical solution requires a selection of the components having the lowest initial cost and the lowest maintenance. The long-range-cost factor is too often overlooked by builders in their concern for how much money must be raised at the beginning of a project.

The selection and evaluation of wall materials is listed below in the order of their importance. The author would certainly include on this list salvage value. Cheaper and easier demolition becomes a significant factor in this era when the average, useful life of a building is comparatively short. Plywood paneling and heavy timber framing have high salvage value. Brick and concrete have low salvage value.

Author's Check List for the Selection of Wall Materials
(In Order of Importance)

1. Initial cost and subsequent maintenance cost
2. Compressive and traverse strength

3. Resistance to natural weathering, chemical attack, and atmospheric pollution
4. Combustibility
5. Ease of handling and erection by size, weight, and shape
6. Resistance to scratching and impact
7. Dimensional changes resulting from temperature and moisture
8. Susceptibility to insect attack
9. Appearance in color and texture
10. Resistance to moisture penetration
11. Sound insulation and absorption
12. Adaptability to future changes of layout
13. Salvage value

Wall materials must have the strength under conditions of compression, bending, shear, and tension to carry applied loads and to resist the external pressure of wind loads. An accurate assessment of the potential strength of a material must be known before one can efficiently and economically design a wall. In the analysis of wall construction, it is important to remember that the loads in question only affect an analysis of the thickness of the wall and do not include any door or window openings. From a design standpoint, there should, therefore, be a greater consolidation of openings, leaving adequately spaced, solid wall panels to provide bracing for the building. Types of bracing and methods of fastening greatly influence the strength and the rigidity of a wall.

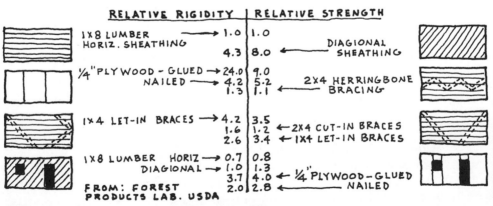

RELATIVE RIGIDITY | RELATIVE STRENGTH

1X8 LUMBER HORIZ. SHEATHING → 1.0 | 1.0
4.3 | 8.0 ← DIAGONAL SHEATHING

¼" PLYWOOD - GLUED → 24.0 | 9.0
NAILED → 4.2 | 5.2 ← 2X4 HERRINGBONE
1.3 | 1.1 ← BRACING

1X4 LET-IN BRACES → 4.2 | 3.5
1.6 | 1.2 ← 2X4 CUT-IN BRACES
2.6 | 3.4 ← 1X4 LET-IN BRACES

1X8 LUMBER HORIZ → 0.7 | 0.8
DIAGONAL → 1.0 | 1.3
3.7 | 4.0 ← ¼" PLYWOOD-GLUED
FROM: FOREST 2.0 | 2.8 ← NAILED
PRODUCTS LAB. USDA

33.1 WALL BRACING TESTS

The durability of a wall is more dependent on heat flow (that is, on temperature change) and on moisture than on any other factors. Moisture deterioration can take place in all organic building materials. Such materials are hygroscopic, absorbing moisture from the air in proportion to the relative humidity. Dimensional change in a wall primarily causes weakness and failure of the wall's fastening devices. Practically all wall materials change their dimensions according to whether they are wet or dry. When a wet material is dried, shrinkage occurs. When the material becomes wet again, expansion takes place. But subsequent expansion is successively less each time, leading to irreversible contraction of the material. Swelling and shrinkage due to moisture changes also cause a breakdown of surface finishes. The pore structure of organic wall materials easily permits capillary moisture attraction from the atmosphere. As much as 40 percent of the dry weight of a material can be taken up and held with water. This is to say nothing of the effect of direct moisture contact from rain or melting snow.

Wall moisture originating from within a building can be an even more damaging factor than outside moisture penetration. As the inside temperature increases, inside water vapor is transmitted into the wall where it condenses. This condensation results in a wetting of the structural materials and in a loss of the insulating qualities of these structural wall materials. It also gives rise to such serious problems as the chemical, physical, or biological deterioration of the wall materials, such as the corrosion of metal, the spalling of brick, and the rotting of timber.

In wood-frame walls, the construction of a vapor barrier resists the movement of the warm, moist interior atmosphere toward the exposed, colder, exterior wood members, where the resulting condensation of water vapor will eventually destroy structural members. Conversely, in masonry walls the use of a vapor barrier, while it prevents moisture from entering the room, also creates a situation where moisture condenses behind the vapor barrier inside the wall, when rain is followed by hot sun, driving moisture inside the wall. Waterproofing the outside of a masonry wall will prevent this moisture migration from the outside inward, but this same waterproofing causes a build-up of water condensation inside the wall. In short, a wall must be designed to limit this outside entry of water via capillarity, while,

33.2 LOCATION OF INSULATION AND VAPOR BARRIERS

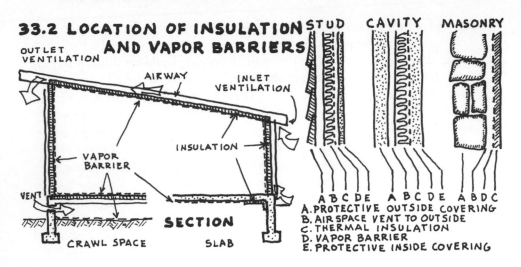

STUD   CAVITY   MASONRY

OUTLET VENTILATION

AIRWAY

INLET VENTILATION

INSULATION

VAPOR BARRIER

VENT

SECTION

CRAWL SPACE   SLAB

A B C D E   A B C D E   A B D C

A. PROTECTIVE OUTSIDE COVERING
B. AIR SPACE VENT TO OUTSIDE
C. THERMAL INSULATION
D. VAPOR BARRIER
E. PROTECTIVE INSIDE COVERING

at the same time, it permits the passage of water vapor from the interior of the wall to the outside. The external wall should have a capacity for the free exchange of air and of water and water vapor to the outside of the wall. Furthermore, the transfer of water vapor to the inside of the wall should be controlled by venting. Obviously, the best choice for a masonry wall is one of the vented-cavity type. Where insulation is used as well as a vapor barrier, a higher indoor humidity may possibly occur without surface condensation. However, incorrect use of thermal insulation can increase the danger of condensation on the interior wall surface, causing paint peeling and other similar damage.

In wall-material literature one finds ample reference to and speculation on the creation of a wonder material which will satisfy all the requirements of an inexpensive, durable, strong, simply constructed wall. Ideally, this material would also exhibit no thermal expansion, no moisture expansion, and it would be impermeable to moisture and resistant to heat flow. Also, there are summer moisture and heat gains and winter heat-loss factors to be balanced out, so it is quite unlikely that such a miracle product will ever exist.

The more practical approach to wall building is for the owner-builder first to plan his walls in relation to areas to be used for winter functions and those to be used for summer functions, those used for

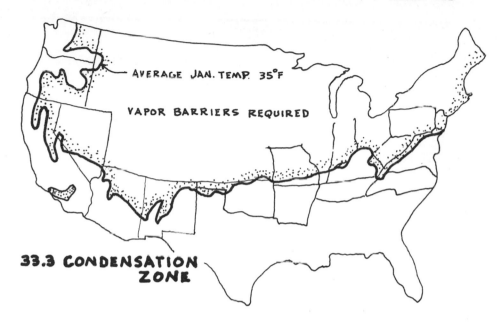

AVERAGE JAN. TEMP. 35°F

VAPOR BARRIERS REQUIRED

33.3 CONDENSATION ZONE

daytime use and those to be used for nighttime use. Walls should be planned as inside partitions and for outside exposure. Bearing walls and curtain walls, storm sides and sunny sides all affect wall planning. Our practice of building all four walls of a house of equal thickness and with equal amounts of insulation is ineffectual. Nighttime sleeping areas have opposite comfort requirements from daytime living areas.

The thermal performance of a wall is determined by (1) its degree of direct, solar-heat penetration through openings, (2) the absorption of solar radiation through the exposed wall surface, (3) heat-storing capacity, (4) insulation characteristics, and (5) ventilation rate. Heavyweight wall materials like tamped earth are cool during the day and warm during the night in regions where the daily variation between outdoor air temperature and solar radiation is great. But, in regions where the daily outdoor temperature variations are small and solar radiation intensities are high, such heavyweight wall materials do not permit sufficient cooling during the night for comfortable sleeping. In these regions, lightweight wall materials should be used.

Another gross extravagance in the home-building industry is the use of heavyweight wall materials in nonload-bearing partitions. This is an example of the builder's failure to select a wall material for its

TABLE 33.1

INSULATION OR RESISTANCE VALUE OF MATERIALS

Material	Thickness in Inches	Resistance Rating
Air space	¾ or wider	.91
Aluminum foil		
Sheet type, one side (with		
¾ or more air space)		2.44
Asbestos board	¼	.13
Blanket insulation	1	3.70
Blanket insulation	2	7.40
Blanket insulation	3	11.10
Brick, common	4	.80
Cinder block	8	1.73
Concrete	10	1.00
Concrete block	8	1.00
Door, wood	1	1.43
Fill insulation		
ground corn cobs, fine	3⅝	8.85
chopped straw	3⅝	5.18
shavings or sawdust	3⅝	8.85
fluffy rock or mineral fiber	3⅝	13.40
Gypsum board	½	.35
Hollow tile, two cell	8	1.67
Insulation board, typical fiber	½	1.52
Insulation board, typical fiber	1	3.03
Plywood	⅜	.47
Roofing, roll, 55 (vapor barrier)	⅛ to ¼	.15
Sheathing and flooring	¾	.92
Shingles, asbestos		.17
wood		.78
Stone	16	1.28
Siding, drop	¾	.94
lap		.78
Window, single glass		.10
Window, double glass		
(in single frame)		1.44

purposeful functioning as well as for its availability and economy. An interior, nonload-bearing wall does not in any way require the properties of an outside, supporting wall.

Wall materials in a house should be as varied as window sizes or roof coverings. The success or failure of a wall design, and of the completed house as well, is determined by the degree to which the builder relates the wall material to the inside and outside environments. A wall is a transition between these two environments. It can express this function either with insensitivity or with a statement of simple logic. When peppered with openings and covered with conflicting materials, a wall lacks clarity and composition. Of all that man has created, simplicity of structure is an enduring characteristic of the finest. One can achieve a unified design by reducing the number of materials used, by consolidating specific areas and spaces. One can simplify linear design by extending lines both horizontally and vertically, resulting in a clean, uncluttered silhouette. Simplicity implies balance in form, scale, texture, and color. Simple structure conveys economy. All this articulation relates to the owner-builder that structure is effective when it satisfies human need. Apply reason to the simplest, most effective means possible for achieving one's goals. Immerse one's self in this economy of means. After a while, one will not have to think effectiveness and balance. It will become part of one's nature; one's actions will respond on a more intuitive, nonverbal, nonintellectual level.

34 WOOD ROOFS

34.1 EVOLUTION OF THE ROOF

After leaving the tree for a drier, safer cave, the first shelter man built consisted of a pit dwelling. This hunter's pit had an A-frame roof of woven twigs over heavier log beams. It combined the shelter of the tree with the protective aspects of the cave, Figure 34.1 (A).

When man later discarded his nomadic ways in favor of animal and plant husbandry, he was forced to raise the ridge of his roof to provide more space for grain-and-tool storage. This caused the roof to sag, so it became necessary to prop the ridge with a center post. This post was symbolic of the tree trunk as the shelter form still familiar to the early farmer whose ancestors had lived the arboreal life, and it was, therefore, given special treatment. On this account, this tree-trunk support post was embellished and honored more than any other timber in the home. This member was called the "king post," a term which survives to this day. See Figure 34.1 (B).

Eventually, man decided to bring under shelter his newly domesticated oxen. It was awkward getting oxen into a pit dwelling, so he made space by placing the roof on walls built above ground. Actually, the first wall was formed when man raised the roof on a timber frame, as noted in Figure 34.1 (C).

The final step in roof evolution took place when our forebears freed the space of the house interior by removing the king post. Structural demands of the newly formed walls required the development of tie beams to hold the walls together, and so the king post was simply elevated to a position where it supported the roof ridge atop the newly installed tie beam. In many old English cottages one still finds reverent care given to the king post. Beautifully carved and decorated, this central prop for the roof harkens to primitive times when tree trunks were much more to man than symbols. See Figure 34.1 (D).

Amazing as it may seem, the king post, truss type of roof had few major improvements prior to research done in 1955 by the University of Illinois Small Homes Council. Recent developments in roofing made by the Virginia Polytechnic Institute have replaced costly, glue-nailed gusset plates with screwtite nails. A relatively large plywood gusset plate at the heel joint permits even greater nailing area and, consequently, even greater joint rigidity. Glue-nailed, king post, trussed-rafter tests made by the U.S. Forest Products Laboratory show that this type of roof provides an ultimate load-carrying capacity of 177 pounds per square foot, or 4.4 times the specified design load of 40 p.s.f., with rafters spaced on 2-foot centers. Substituting threaded nails for glue made possible even greater failure loads of 5.5 to 7 times the design load. These phenomenal structural results, which can carry spans of up to 32 feet, require rafters no stronger than common 2 by 4s.

Today, a wide variety of trussed rafter systems are being used throughout the world, and interesting variations of basic systems appear from widespread sources. Builders everywhere readily recognize the advantages of trussed rafter construction. For one thing, individual roof trusses are precut and preassembled on the ground, usually on a table-height jig for greater speed and accuracy. Only 70 board feet of small-dimension lumber are required on one trussed rafter spanning 32 feet. A trussed roof can be covered for weather protection in record time. Large spans bearing on outside walls enable

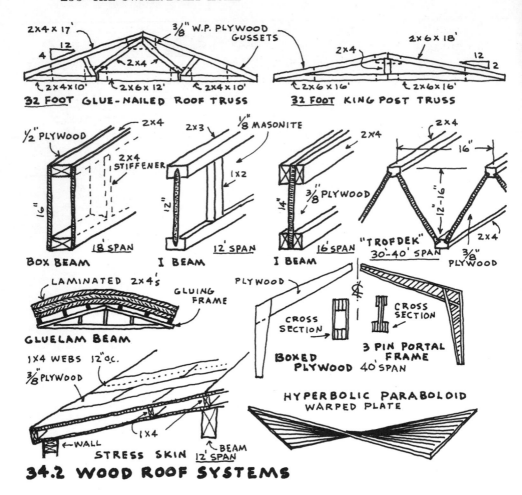

34.2 WOOD ROOF SYSTEMS

later fabrication of nonload-bearing, light partition walls. Later planning changes and remodeling are, thereby, made possible.

The plank-and-beam roof has already been discussed in chapter twenty-five. In some respects this system is the most suited to owner-building. Fewer but heavier roof members are used, and the exposed beam roof combines structure with finish in one operation.

To design a plank-and-beam roof strong enough to safely carry the total roof load and stiff enough to prevent roof sag, one must first take into account the weight of the roof (the dead load) and the weight

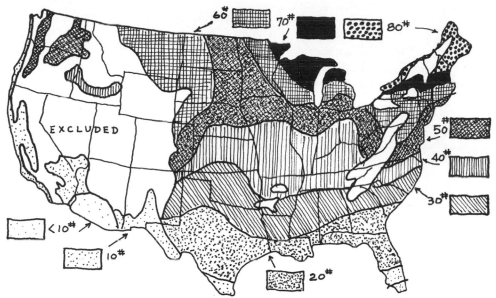

34.3 MAXIMUM SNOW LOADS IN POUNDS PER SQUARE FOOT

of maximum snow loads and other live loads. Also very important to the design of a post-and-beam roof system are limitations which result in load deflection (roof sag) and which result from variations in the structural properties of the wood to be used. These variations include fiber stress, tension, and the compression developed in sagging (deflection) under total load. A horizontal shear stress is also present in a roof beam under load. This shear stress is the strain resulting from the tendency of two adjacent parts of the internal structure of the wood beam to slide on each other, with maximum horizontal shear stress occurring midway between the top and bottom of the beam. Lumber is used to its maximum efficiency when deflection and bending stresses are simultaneously satisfied. Increasing the beam width without increasing the spacing of the span may satisfy the demands of horizontal shear, but it will leave unused a portion of the potential strength of the wood beam to satisfy deflection requirements. The owner-builder should consult beam-design charts published by the National Lumber Manufacturers Association.

Trusses and roof beams are roof systems which use individual pieces of timber in an unmodified form. The joint is usually the weakest point in this type of roof. For this reason and for purposes of developing a type of roof system where a high ratio of strength-to-weight exists, new solutions—highly adapted to owner-built construction—have recently appeared. The box beam is one solution to this current demand for more strength, using less material. Being hollow, deep-section box beams are light and easily handled with up to as much as 40-foot spans. "I" section beams have a plywood web, which transmits tensile stresses to paired flanges designed to withstand all bending stresses.

Casein glues are commonly used to bond laminations of parallel-grain wood. Thus, little warping occurs and higher design stresses can be used, since individual laminations can be graded separately and higher-grade pieces can be positioned where the highest stresses occur. Furthermore, laminated members can more readily be formed into curves, and sectional dimensions can be increased at points of high stress or be reduced at points of low stress.

As already mentioned in chapter twenty-five, stressed-skin panels make an ideal solution to practically any roof problem. Thin plywood sheets form structural skins, which are glue-nailed to stabilizing webs, forming an enclosed roof panel. The roof panels function as a network of I beams, their skin acts as a flange overcoming axial bending stresses, and their webs act in shear to stabilize the skins against buckling. A typical panel is made with 1- by -4-inch wood webs and $\frac{3}{8}$-inch plywood skins, spanning 12 feet under 50-pound snow loads. Modified panels can be made to span 16 feet, thereby eliminating rafters and joists and combining the roof and ceiling in one unit.

Curved, stressed-skin roof construction provides maximum strength-to-weight ratios. Stresses in a curved skin are transferred equally to boundary-beam members, providing the greatest clear floor area with the minimum number of supporting points. The roof shell can consist of two or three cross laminations of boards or plywood, glue-nailed together to form the required profile. Boundary and intermediate beams can be of glu-lam (glue-laminated) construction. More mathematical calculation, site programming, and fabrication detailing are required with glu-lam construction as compared with conventional rafter-joist construction, but the owner-builder, having more time than cash, is responsive to these material-saving solutions.

35 MASONRY ROOFS

Anyone who has done much world traveling has no doubt observed an interesting housing paradox: both indigenous people and wealthy people tend to have the most effective housing for their respective means. The worst housing seems to be found in working-class industrial slums and in middle-class development tracts, where contractor-speculator-built houses incorporate shoddy design, construction, and materials. Primitive, indigenous housing, on the other hand, invariably has the best climate-attuned design features and the most sensible use of native materials. An exceptionally good wall material like adobe or tamped earth can be afforded by the penniless poor, because earth is a freely available material, requiring only labor to form it structurally. But only in upper-class neighborhoods does one find expensive California dobe.

Free labor (and all too often free earth) is one commodity which the middle-class, nine-to-five city worker or the industrial-slum worker does not have. Therefore, he must rely on inferior building means, such as building with plaster on studs. Commercially produced adobe block walls and monolithic earth walls are second in cost only to stone. One finds earth walls either in the most expensive homes—or in the humblest hut.

A second example of this housing paradox, and one relative to the subject of roofing, is precisely illustrated in a book, *Soil Construction,* issued in 1956 by the Ministry of Labor of the State of Israel. A hyperbolic-paraboloid roof shape, heretofore reserved for the most expensive, conversation-piece architecture, has been developed by the Israeli government for experimental, low-cost housing. All the numerous advantages of the thin-shell, reinforced-concrete roof can now be profitably utilized by those in the lowest economic brackets of world society. Significantly, earth block was the wall material chosen by the highly qualified Israeli engineers who developed this low-cost project.

The hyperbolic paraboloid was first used shortly after World War II. Being a relatively simple shell to design and to construct, its use spread widely. Its interesting warped shape is produced by straight-line segments. In addition to saving materials, this type of roof is strong and rigid. A two-way curve limits the possibility of cracks resulting from shrinkage due to temperature changes. The pavilion type of shell roof can be erected on square or rectangular buildings. Each roof consists of four hyperbolic-paraboloidal surfaces, each bounded by two horizontal lines and two sloping lines. These surfaces intersect along horizontal lines, forming a cross, which divides the roof into four equal quarters, while the edges of the roof are sloping.

Soon after the results of the Israeli research project were made available, I built a 16-foot-square drafting studio, incorporating similar principles of structural design. The 2-inch-thick concrete roof to this day remains crack free, maintenance free, and fully fireproof. As a matter of fact, the money saved from not having to pay fire-insurance premiums over the past twenty years has more than paid for all the materials that built the studio and the attached residence, which is, likewise, of masonry construction.

The numerous advantages of an all-masonry roof are available to rich and poor alike. Of the wide variety of masonry-roof systems discussed in this section, perhaps the oldest is that which employs a composite of tile block and concrete. Ancient Romans used burnt-clay tile extensively in their roof constructions, and, still today, Italy leads the world in the use of clay-tile building units. Tile blocks are placed on solid formwork, and concrete ribs are poured between the two. These are called two-way spans. One-way tile spans can be made which do not require this expensive formwork, but they have less

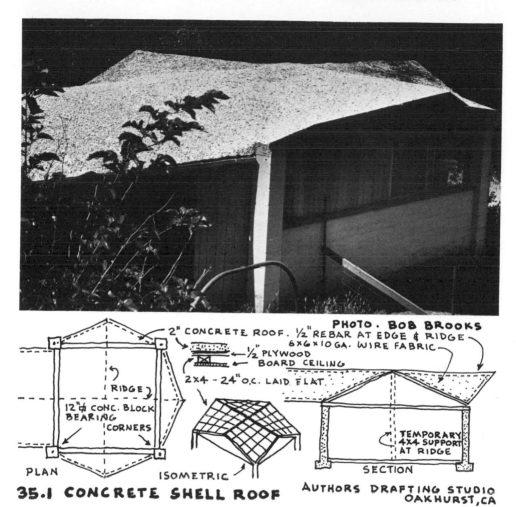

PHOTO · BOB BROOKS

2" CONCRETE ROOF. ½" REBAR AT EDGE & RIDGE
6×6×10 GA. WIRE FABRIC
½" PLYWOOD
BOARD CEILING
2×4 - 24" O.C. LAID FLAT

RIDGE

12"∅ CONC. BLOCK
BEARING
CORNERS

TEMPORARY
4×4 SUPPORT
AT RIDGE

PLAN   ISOMETRIC   SECTION

35.1 CONCRETE SHELL ROOF   AUTHORS DRAFTING STUDIO
                             OAKHURST, CA

strength. The two-way slab consists of a square or nearly square panel
having beams on either side. Two sets of reinforcing bars are located
at right angles to each other, thereby distributing the load to each
of the four beams. Less steel reinforcement and concrete is needed
in a two-way slab, as a certain degree of arch effect is realized. Also,
the weight of a tile roof is less than that of a monolithic concrete
roof. The lightweight tile roof makes possible a use of smaller beams,
columns, and foundation.

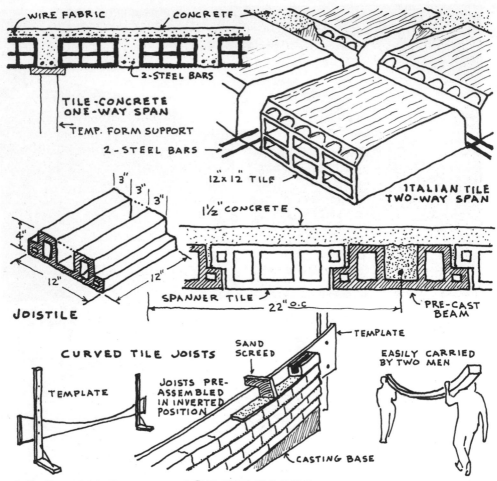

WIRE FABRIC   CONCRETE
2-STEEL BARS
TILE-CONCRETE ONE-WAY SPAN
TEMP. FORM SUPPORT
2-STEEL BARS
12x12" TILE
ITALIAN TILE TWO-WAY SPAN
3" 3" 3"
4"
12"   12"
1½" CONCRETE
SPANNER TILE   22" o.c
PRE-CAST BEAM
JOISTILE
CURVED TILE JOISTS
SAND SCREEN
TEMPLATE
TEMPLATE
JOISTS PRE-ASSEMBLED IN INVERTED POSITION
EASILY CARRIED BY TWO MEN
CASTING BASE

35.2 TILE-CONCRETE ROOFS

In 1938, Iowa State College developed the first precast tile-concrete beam. A specially designed spanner tile was placed between beams, and concrete was poured over the entire surface. It was found that the bond between the concrete and the tile was sufficient to cause the two materials to act as a unit. Later, tests at the University of Texas showed that the greater tensile strength of tile increases the diagonal tension or the shear strength of tile-concrete beams over equivalent plain concrete beams.

The most recent tile-concrete roof system, an improvement over its forerunners, is called "Joistile." The spanner tile and beam tile of this roof are of the same design, reducing production costs. To build a Joistile roof, one first lines up beam tile, end to end, on the ground until the desired length is obtained. Steel is placed in the trough which is filled with concrete. After curing, the tile-concrete beams can be hoisted into position on the roof. Filler tiles are then placed between beams, and a topping of concrete is poured over the whole surface. Tile-concrete roof spans should be less than 20 feet. Installation costs of from 50 to 60 cents per square foot are competitive with other types of roofing, such as wood-framing systems.

The preformed, tile-concrete joist system lends itself nicely to parabolic or circular, arched-roof construction. First, one need only build a template to the desired curvature, and then one preassembles the tile on the template in an inverted position. After the concrete-bonding reinforcement rods have set and cured, the joists can be easily handled by two men.

In regions where tile is not readily available, one can substitute

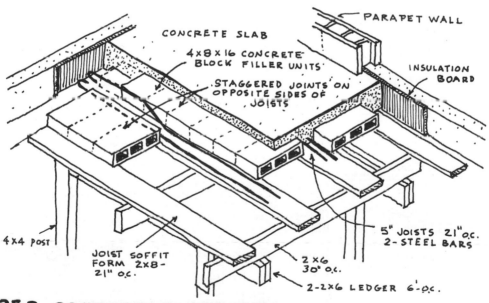

35.3 CONCRETE BLOCK ROOF

concrete block with equally good structural results. Like tile, the concrete block serves the dual function of form for the beam and for a flat ceiling, which can either be plastered or painted. Fewer open-deck forms are required when soffit-type, concrete-block filler units are used. Also, the flat ceiling becomes more uniform, for no beam line is exposed on the underside.

Where no major obstacles are met in the forming of the entire roof area, one can economically pour a combination insulation-joist-slab roof. First, tapered wood joists are set on the formwork at about two-foot intervals. Lightweight concrete is poured between the wood joists. The wood members are then removed and reinforcing bars are placed in the grooves. These ribs are then poured along with a thin roof slab, which is poured over the insulation layer.

In 1968, the Government Building Research Station at Lahore, West Pakistan, developed a tile-roof system which turned out to have amazing structural qualities. Each tile weighs 70 pounds and is only 1 inch thick, but, in place and covered with 4 inches of soil-and-mud plaster,

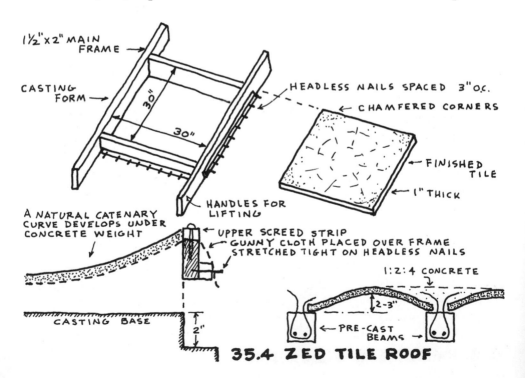

35.4 ZED TILE ROOF

the roof supports over 300 pounds per square foot. These tiles, called Zed tiles, are cast on a raised platform on gunny cloth in a 30-inch-square frame. Soon after pouring the concrete, the frame is lifted off the platform and placed on four brick corner supports. Naturally, the wet tiles sag under their own weight, producing a catenary curve. In a half-hour, the upper screed strip can be removed, and, after one day, the tile can be removed from the gunny cloth form. After the Zed tiles are cured in a water tank for 3 to 4 weeks, they can be placed between precast concrete joists and covered with concrete.

Often, it proves advantageous to precast concrete joists on the ground. For one thing, less formwork is required. Welded-wire fabric, which is attached to strong, water-resistant paper backing, can be used as a combination reinforcement and forming material over precast joists. Although the heavy paper conveniently acts as a form for the concrete, it unfortunately prevents the concrete slab from bonding with the joist, thereby preventing a T-beam action.

Welded-wire fabric is the most essential reinforcement for concrete roofs. It is easily placed in position by draping it across joists. Negative bending moment near the joist supports and positive bending moment at midspan are resisted by this reinforcement, and, therefore, minimum concrete cracking results.

The usual precast concrete joists resembles an I section. All reinforcing bars are assembled in advance to form a reinforced cage. When the reactions of a joist under load are analyzed, one can readily understand why each piece of reinforcing is positioned as it is. Molecules above the neutral surface of a joist are in compression, whereas, below the joist, they are in tension. There also exists a diagonal tension or shear force, which causes vertical cracks at the center of the span and cracks at an angle of about 45 degrees at each support. Vertical, reinforcing rods, called stirrups, are usually welded or wired to the top and bottom moment rods. Hooks at the end of U-shaped stirrups provide anchorage for the resistance of tensile stresses.

Unfortunately, a concrete roof expands and contracts as it warms and cools, resulting in troublesome cracking. Also, a roof is subjected to low outside temperatures simultaneously with warm inside temperatures during the winter months and to reverse temperatures during the summer months. In this order, a combination of vapor barrier, insulation, waterproofing, and reflective roof surfacing placed above the concrete roof-deck will minimize all the air, moisture, and heat problems which confront the user of concrete roofdecks.

As mentioned earlier, roof insulation also increases the likelihood of condensation because of a reduction in the surface temperature on the cool underneath side. Furthermore, outside waterproofing tends to further trap water vapor in the slab. A ventilated air space between the insulation and the roof slab will alleviate this problem.

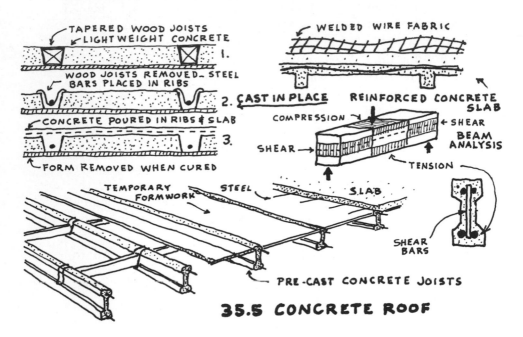

35.5 CONCRETE ROOF

The best choice for concrete decks is built-up roofing surface. This type of roofing surface is built up, on the job, with three or four layers of tar-saturated felt placed between applications of hot asphalt. Preferably, the finished, gravel surface should consist of dolomite or white-marble chips. For steeper roofs, a whitening treatment has been developed by the British Building Research Station. A mixture of quicklime and tallow will not deteriorate the asphalt undercoating as most other coverings will do. The mixture is prepared by, first, breaking up a white, high-calcium quicklime into small lumps and by adding 10 percent tallow by weight, which is shredded and placed on top of the heap. Sufficient cold water is then added so that the heat developed in slaking is adequate to melt the tallow and to disperse it within the mass, without its becoming charred and destroyed. When slaking is complete, sufficient water is added to permit the mass to be worked into a stiff cream.

My travels in Norway have brought to my attention the most effective roof-covering material that is available—old mother earth, or sod. Coincidentally, only the peasant cottages retain this use of sod roof coverings for their dwellings. A mere 5 inches of pasture dirt is applied to a rough-laid sheathing sealer of birch bark. There is no better example of how a native material can be utilized by an indigenous people to their practical advantage. In winter, the dead grass covering holds snow, providing additional insulation. In the spring, rains beat down the resilient, new-growth grasses so that, thatchlike, they quickly shed excess water. In the summer, the full-grown grasses create air movement about their tall stems, effectively insulating the roof and even reflecting the sun's heat.

The owner-builder might prefer a somewhat thicker sod covering; say, 8 inches. A built-up underlayer of hot tar and fiberglass roofing would probably also be preferable, for its durability, as compared with a birch-bark underlayer. Or, if a cold roofing application is preferred, one can embed a layer of 6 mil black polyethelene plastic into black mastic over a layer of 45-pound mica-surfaced roofing felt.

36 STAIRS

The National Safety Council claims that the greatest number of home accidents occur on stairways. For this reason there are a few common-sense rules that the owner-builder might keep in mind when stairways are planned. Consider that:

1. The maximum height of a riser should be 8 inches
2. The minimum tread width should be 9 inches, excluding nosing, which should not exceed $1\frac{3}{4}$ inches
3. The height of a riser plus the width of a tread should equal not less than 17 inches nor more than 18 inches—$17\frac{1}{2}$ inches is considered ideal for this measurement
4. All risers and treads in the same flight should be equal
5. Landing dimensions should be equal to or greater than the stairway width
6. All stairways should have a permanent handrail, 36 inches in height from the center of the tread
7. Headroom between the front edge of the step and the nearest vertical obstruction should not be less than 6' 6"
8. Minimum stairway width should be 3 feet

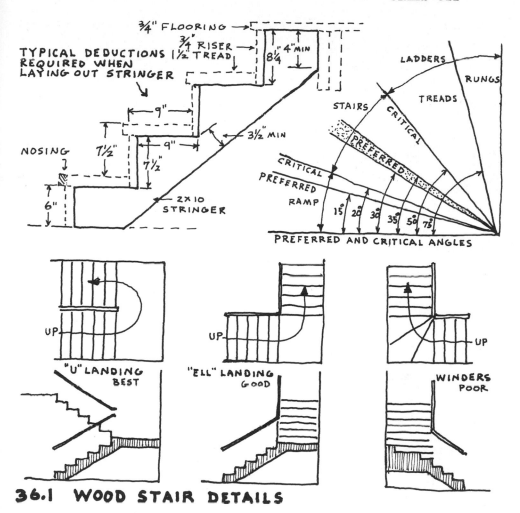

36.1 WOOD STAIR DETAILS

The simplest type of stairway to build is the straight-flight stair. Wood is the least costly material to use for stringers and steps. Steps are notched out of a pair of 2- by -10-inch or 2- by -12-inch stringers. At least $3\frac{1}{2}$ inches should be left in the stringer after step cuts are made. Also, adequate bearing (at least 4 inches) must be provided against the header, so that the stringer can be securely fastened.

The collaborating engineers of Colombia, South America, who invented the Cinva block-making press also developed a prefabricated,

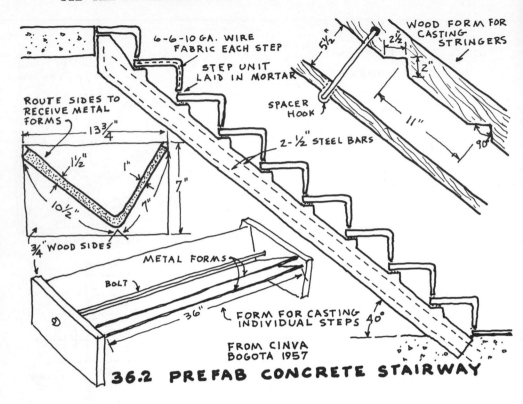

6-6-10 GA. WIRE FABRIC EACH STEP

STEP UNIT LAID IN MORTAR

WOOD FORM FOR CASTING STRINGERS

ROUTE SIDES TO RECEIVE METAL FORMS

13¾"

SPACER HOOK

2-½" STEEL BARS

5½"

2½"

2"

11"

90°

1½"

1"

10½"

7"

7"

¾" WOOD SIDES

METAL FORMS

BOLT

36"

FORM FOR CASTING INDIVIDUAL STEPS

40°

FROM CINVA BOGOTA 1957

36.2 PREFAB CONCRETE STAIRWAY

concrete stringer-and-step system, which is ideally suited to low-cost owner-builder construction. The riser-tread units are cast in a metal mold, and, when cured, they are laid in mortar on the reinforced concrete stringers.

Special room-planning or traffic-circulation problems may dictate some other stair arrangement besides the simple straight-flight stairway. As illustrated here, an alternative of a U-landing, an L-landing, winders, or spiral-stair system can be built.

In 1969, Forestry Sciences Laboratory of the Department of Agriculture designed an interesting, low-cost, spiral-stairway building system. A ¾-inch steel rod through the center pole supports, in tension, all the treads and riser blocks of the spiral stairway. Spiral stairs are great space savers and are especially attractive for access to sleeping lofts.

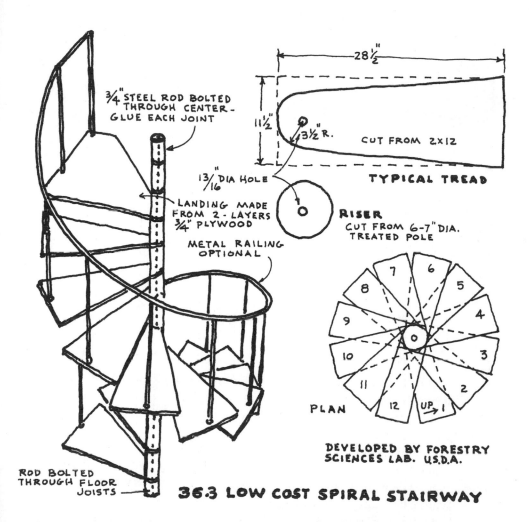

3/4" STEEL ROD BOLTED THROUGH CENTER- GLUE EACH JOINT

28 1/2"

11 1/2"

3 1/2" R.

CUT FROM 2x12

TYPICAL TREAD

13/16" DIA HOLE

LANDING MADE FROM 2 - LAYERS 3/4" PLYWOOD

RISER
CUT FROM 6-7" DIA. TREATED POLE

METAL RAILING OPTIONAL

7 6 8 5 9 4 10 3 11 2 12 UP 1

PLAN

DEVELOPED BY FORESTRY SCIENCES LAB. U.S.D.A.

ROD BOLTED THROUGH FLOOR JOISTS

36.3 LOW COST SPIRAL STAIRWAY

Whenever possible I prefer to incorporate upper-story stairs into the fireplace-masonry core. The centralized fireplace is a logical point of access to either upstairs loft-sleeping or to downstairs basement storage. Also the fireplace support offers excellent vertical support for a cantilevered, masonry stair tread.

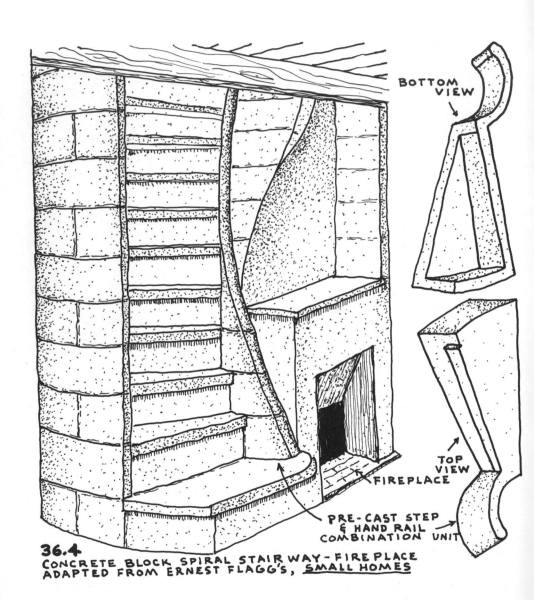

BOTTOM VIEW

TOP VIEW

FIREPLACE

PRE-CAST STEP
& HAND RAIL
COMBINATION UNIT

36.4 CONCRETE BLOCK SPIRAL STAIRWAY - FIREPLACE
ADAPTED FROM ERNEST FLAGG'S, SMALL HOMES

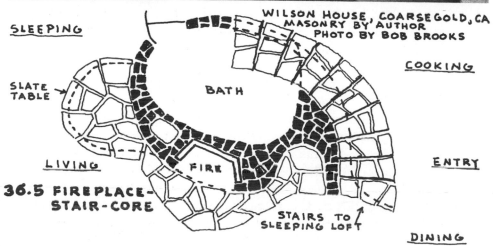

WILSON HOUSE, COARSEGOLD, CA
MASONRY BY AUTHOR
PHOTO BY BOB BROOKS

SLEEPING

COOKING

SLATE TABLE

BATH

LIVING

FIRE

ENTRY

36.5 FIREPLACE-STAIR-CORE

STAIRS TO SLEEPING LOFT

DINING

37 PLUMBING

Modern architecture is a revolt against styles and is based on the intimate awareness of functional requirements in the broadest sense of the word "function." Unfortunately, the revolt preceded the research needed to start establishing these functions.

—*The Architects Journal*, 1965

Far too many owner-builders warily resist the idea of installing their own plumbing and wiring. These amateurs somehow blunder through design and structural problems but retreat for comfort and reassurance in favor of experts when the time comes to plumb and to wire their houses. The duly licensed experts are, therefore, apt to hoodwink their clients into believing that their profession, above all others, requires special aptitudes and near-magical powers of understanding and execution.

True, when installed under building-code jurisdiction, the variety of pipes and fittings and the usual complexity of vents and drains is enough to exhaust the hardiest novitiate. But plumbing-fixture installments can be simplified. New fittings and pipe materials and improved drain and vent layouts are now available. A rational understanding of plumbing practice exposes to scrutiny much of the costly hocuspocus to be found in the secret order of the plumbing trades.

Officials of this trade covetously protect their profession from renegade intruders and owner-builders by unfurling the plumbing code. The only really just part of the U.S. Plumbing Code is section 1.10 (c) which states:

Any permit required by this code may be issued to any person
to do any plumbing or drainage work regulated by this code in
a single-family dwelling used exclusively for living purposes . . .
in the event that any such person is the bona-fide owner of any
such dwelling, and that the same are occupied by or designed to
be occupied by said owner, provided that said owner shall personally purchase all materials and shall personally perform all labor
in connection therewith.

Plumbing codes, now in force, date back to the 1870s when the
water-carriage system was first introduced. The erroneous nineteenth-century belief that sewer air causes disease is still perpetuated
by law. Sewer-air code requirements require individual venting for
each fixture trap. As background information about the function of
vents, one may be advised that the fixture trap provides a water seal
between the drainage system and the building's interior. When a
fixture is discharged, a pressure fluctuation occurs which tends to
reduce the water seal in the fixture trap. Therefore, vent pipes are
connected at various points in the drainage system so that gases may
escape to the outside, thereby reducing the pressure at the fixture.
Those plumbing officials who continue to hold the discredited, sewer-
gas theory would do well to read Winslow's *The Sanitary Significance
of Bacteria in Air of Drains and Sewers*. According to Winslow, as
quoted by the American Public Health Service, a person who placed
his mouth atop a plumbing stack and breathed the air from it for 24
hours would inhale no more colon bacilli than were then to be found
in a quart of New York City drinking water.

If a break in the operation of a fixture trap occurs, its only effect
is a musty odor, immediately corrected by refilling the trap. A separate
vent pipe for each fixture is not justified on health grounds. Self-
siphonage, the discharge from a fixture sucking away its own trap seal,
is more likely with bowl-shaped than with flat-base fixtures for the
final, slow trickle of the latter insures a refilling of the trap.

Functionally, one 2-inch stack can well serve the total venting
requirements for a complete house-plumbing system. Also, this stack
can be used to carry discharge from fixtures connected into the drainage system at a higher level. This is called wet venting. However,
according to code requirements, the one-stack plumbing layout is only

permissible when the fixture drain slopes continuously at $\frac{1}{4}$ inch per foot, and when the length of drain between trap and vent does not exceed the following: 5 feet for $1\frac{1}{4}$-inch drain, 6 feet for $1\frac{1}{2}$-inch drain, 8 feet for 2-inch drain, and 12 feet for a 3-inch drain. Proper fixture arrangements, therefore, yield substantial material and labor savings.

The number one rule in plumbing-fixture arrangement is *keep fixture placement compact.* Whenever possible, locate the toilet between the tub and the sink, so that the vent for these fixtures can go right up from the toilet fixture. For ease of installation and a saving on materials, bathroom fixtures should line up on one wall. The opposite side of the so-called wet wall could, perhaps, include food preparation or laundry functions. A complete mechanical core might well be planned to include water heater, furnace, fireplace, and the main electrical panel, as well as the bathroom, laundry, kitchen plumbing, and wiring lines. Some progressive plumbing designers have arranged plumbing fixtures so that all supply and drainage pipes are above the floor. This is possible now that the much improved wall-hung toilet is available. With this arrangement, plumbing can be one of the final operations in house construction. Even the fixtures are hung before the pipe is installed. If desired, access to the complete bathroom-plumbing system can be made from behind kitchen counters located on the opposite side of the wall. Above-the-floor plumbing also makes possible the use of less expensive plastic and galvanized-iron drain and waste fittings.

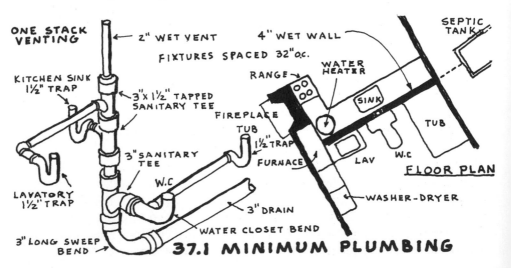

37.1 MINIMUM PLUMBING

The recent introduction of plastic DWV (drain, waste, and vent) pipe and fittings marks one of the few major advances in the plumbing industry since 1920, when the outhouse was moved indoors. This material is made from a black acrylonitrile compound, which is proven to have excellent dimensional stability and resistance to acids and alkalies. Plastic DWV eliminates the usual need for leading, caulking, threading, or soldering joints. It is cut with a carpenter's saw and fitted with solvent cement. In comparing the installed cost of cast-iron, copper, and plastic plumbing, one realizes an average saving of $100 by using plastic DWV. The material cost of cast iron in one typical residential system is $200, with a required 22 man hours of labor, or a total of $400. The cost of copper pipe and fittings is $250 with 12 man hours of labor required, or a total of $350. The material cost of plastic DWV is $150 with 4 man hours of labor, or a total of $200.

Plumbing materials and plumbing techniques have greatly improved in recent years. The same good thinking should now be applied to the design of the bathroom, to bathroom fixtures, and to waste disposal. A proper toilet, for instance, implies a proper disposal system which, in turn, further implies a proper structure to accommodate these facilities.

The criteria for "proper" are: Is it healthful? Is it economical? Is it pleasing to use and to behold? One need not be reminded that modern bathroom systems are ugly, expensive, and unhealthful. If one wishes to know how much a conventional bathroom plumbing-and-drainage system costs, count the number of fixtures in the house, including lavatories, bathtub, and water heater. This total multiplied by $400 gives a close estimate of the cost of this system. Add to this the room area of one's bathroom multiplied by $15 and one has the surprising total of about $3,500 for an average-size bathroom and drainage system.

It can also be reasonably stated that the conventional toilet is not healthful. The high sitting position is artificial and unhygienic. It is likewise unhealthful to pollute bodies of water with water-borne sewage. Each year 4.5 million tons of sewage sludge are dumped into the oceans surrounding North America. This represents 4.5 million tons of nitrite contaminant, and it also represents 4.5 million tons of potentially valuable fertilizer which is not returned to the land.

In his book *The Bathroom,* Alexander Kira quotes medical testimony condemning the seat-height toilet. The modern toilet has been called "the most atrocious institution, hygienically, of civilized life." Health-conscious owner-builders should certainly consider the merits of installing a seatless, squat-type toilet. American Standard, one leading manufacturer of bathroom fixtures, sells a ceramic squat plate through their manufacturing outlet in France. A California plastics manufacturer also sells a fiberglass squat plate patterned after a design developed by the Ministry of Health of Thailand. Interested readers may write Owner-Builder Publications in Oakhurst, California, for details. The finished squat-plate toilet bowl requires about one quart of water for flushing, compared with the 5 gallons required to flush conventional toilets. The bowl can be maintained, quite clean and sanitary, without difficulty, and, most important of all, the use of this bowl makes possible a posture for natural evacuation.

While the one-quart flush is an important water-conservation measure, it also makes possible through composting action, a totally new type of waste disposal. One good summary of all of the various fixtures for and methods of compost-waste disposal is presented in the Minimum-Cost Housing Group's booklet, *Stop the Five-Gallon Flush!,* published by the School of Architecture at McGill University.

There are two, basic methods of excreta disposal, which can be built by the owner-builder: the compost privy and the septic tank. The first process is aerobic and requires oxygen for decomposition in its fermentation process. The anaerobic process consists of putrefactive breakdown in places where oxygen does not have access. We must choose between the processes of fermentation and putrefaction in our attempt to reclaim the nutrient fertilizer value of waste, and to dispose of excreta in a sanitary manner.

World Health Organization publications present a compelling argument against disposing of excreta through anaerobic decomposition in sewage-treatment plants or in septic tanks. In putrefactive action, as a result of oxygen not being/present in the process, no heat build-up occurs and, therefore, certain pathogens and parasites are not fully destroyed. It has been found that contaminated material in liquid suspension, such as that in anaerobic digestion, can remain viable for as long as six months. Another thing in favor of aerobic decomposition is that there are many more species of bacteria involved in aerobic fermentation than in anaerobic putrifaction.

Other problems are associated with the disposal of water-borne waste. Sewage, borne on large quantities of water necessary for its transport, is difficult to treat. Water does have a certain ability for self-purification, but this requires oxidation, and usually the volume of water is too small in proportion to the volume of sewage to supply the required quantity of oxygen. Consequently, water receiving waste becomes foul and water-living flora and fauna, especially those which require oxygen, are destroyed. The water also becomes contaminated with pathogenic bacteria, called protozoa, and with the eggs and larvae of harmful helminths, also known as liver flukes.

Our society not only legalizes pollutive, unsanitary disposal methods, but it also outlaws the eventual return of composted excreta waste to agricultural lands as plant nutrients, which could well be utilized in the prevention of human disease and in the improvement of human nutrition.

The only practical way to reclaim excreta waste is through aerobic composting. Pathogenic bacteria and worm eggs can survive only from 30 minutes to 1 hour in an oxygenated compost environment. Compost temperatures rise to 160°F. High temperatures, however, are only partly responsible for this destruction of bacteria. Competing bacterial flora and predatory protozoa contribute to the destruction as well. Aerobic composting is achieved by a succession of bacterial and fungal populations, each suited to its own environment and to its own relative duration. The activities of one group complement the next.

Humus is the end product of properly composted, organic materials. The production of humus contributes to increased nitrogen fixation in the soil from nitrogen in the air. As the gradual decomposition of insoluble organic matter takes place, nitrogen is liberated as ammonia and is then oxidized into nitrates in the humus. Plants can utilize this nitrogen only in the form of nitrates. So, when raw, uncomposted wastes are spread on the land, as is commonly done in the Orient, nitrogen evaporates into the air instead of being used by plants.

The Indian Council of Agricultural Research at Bangalore began extensive composting programs based on the compost-privy principle. They built an experimental, double vault latrine, each vault with a separate seat. During the time that one compartment was being used, compost material in the adjacent compartment was ripening. A period of six months lapsed between clean outs. This two-compartment system appears to be superior to others. However, by incorporating a simple damper mechanism, only one squatting plate needs to be installed.

Rikard Lindstrom of Tyreso, Sweden, has patented a simple aerobic composting chamber. Its salient feature consists of a sloping 16° bottom on a tank chamber, thereby providing continual movement of decomposing refuse to the far end of the chamber as additional wastes are added. The chamber bottom should contain a 12-inch-thick layer of straw or sawdust, so that urine will be absorbed and may be reclaimed. This porous layer of cellulose also provides aeration of the central section of the pile. Lindstrom used a system of inverted, U-shaped conduits and ventilation holes to provide adequate aeration. Air circulation may also be accelerated through solar-heated flue conversion.

My experimental compost privy is schematically illustrated in Figure 37.2, and Figure 8.6 shows how it might be incorporated with a sun-pit greenhouse. Even though the compost privy has proven to be of merit, its widespread use is not to be expected. There are many social, legal, and technical difficulties associated with the adoption of this new, functional mode of processing human excreta. For clarification of specific aspects of its building and its use, ask your friendly local building inspector. In my judgment the long-term, personal rewards and the benefits to the environment warrant whatever manner of subterfuge deemed necessary to build the compost privy for one's own use. No county building department should have the power to prevent one from squatting to relieve one's self, nor should it prevent compost activities limited to one's own garden.

Traditionally, bathing is considered a ritual of serenity for Japanese, but to Americans it is too often a task to be done as quickly as possible. The innovation of a seat for bathing, to be found on the side of the square tub, is a pleasant improvement over conventional tubs. Built-in, sunken tub-shower combinations offer opportunities for yet additional bathing comfort and aesthetics. Japanese usually furnish bathers with a *furo*, a three-foot-deep tub for hot water, placed next to the shower. The shower is first used to clean one's body scrupulously. Then, one

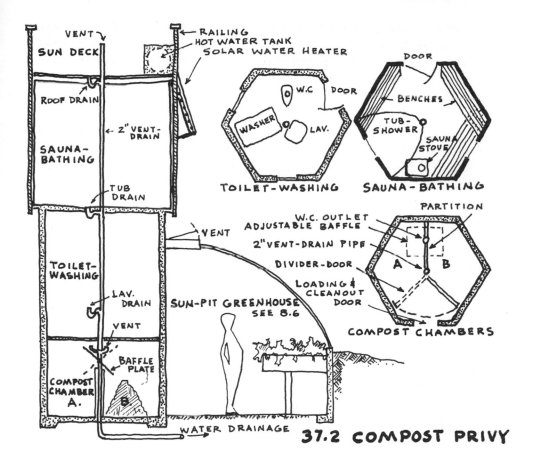

37.2 COMPOST PRIVY

submerges the body into 104° *furo* temperature, experiencing complete physical and mental repose.

An equally ancient method of heat bathing is found in the Finnish sauna. The sauna consists of an air-tight room that is at least 8 by 6 feet in size, and which is heated to a temperature of 175 to 200° with a humidity of 8 percent. This dry heat tends to open and to clean the body's pores of perspiration, while the circulation is stimulated. Stones are heated on top of a sauna stove. A cupful of water is then poured over the stones to provide a quick burst of added heat. The walls and ceiling of a sauna should be lined with unfinished redwood or cedar wood, to better absorb the moisture evaporating from the bathers' bodies and to thereby keep the air freer of moisture and musty odor. Health seekers should definitely install a sauna adjacent to the bathroom shower. Ten minutes in a sauna, followed by a spine-tingling, cool shower, invigorates the body as no other therapy does.

When a direct connection to an established sewer system is not possible, and, when for some reason or other the compost chamber is not an attractive alternative, the owner-builder must provide himself with a water-carriage waste-disposal system—one in which sewage and liquid wastes are conveyed underground by the flow of water. The septic tank and the accompanying absorption field form the two parts of a water-carriage disposal method. There are countless varieties of septic systems, and, before the Public Health Service Environmental Health Center at Cincinnati, Ohio, issued the results of its thorough investigation of the subject, one was very much at a loss to know which household disposal layout to choose.

A septic tank functions in three ways: as a sewage-settling tank, as a sludge-storage tank, and as a digestion tank. Digestion of sewage in a well-designed tank is fairly complete. Anaerobic bacterial action,

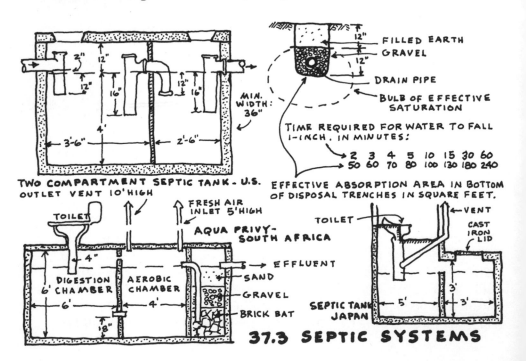

TWO COMPARTMENT SEPTIC TANK - U.S.
OUTLET VENT 10' HIGH

FILLED EARTH
GRAVEL
DRAIN PIPE
BULB OF EFFECTIVE SATURATION

TIME REQUIRED FOR WATER TO FALL 1-INCH, IN MINUTES:
2 3 4 5 10 15 30 60
50 60 70 80 100 130 180 240

EFFECTIVE ABSORPTION AREA IN BOTTOM OF DISPOSAL TRENCHES IN SQUARE FEET.

AQUA PRIVY - SOUTH AFRICA

37.3 SEPTIC SYSTEMS

resulting in the putrefication of the sewage, takes place first. The septic tank should be sufficiently large to permit storage of solids, since digestive action by bacteria is relatively slow. At the inlet end of the septic tank, a baffle or pipe fitting should be installed to divert the incoming sewage downward, leaving the upper scum layer of the tank undisturbed. The outlet end of the septic tank has a similar baffle, which retains solids but permits liquid effluent to be discharged into the absorption field. Initial research at Cincinnati shows that the two-compartment tank results in more efficiency than a one- or a three-compartment tank. Most of the upper-level scum and the bottom-level sludge are retained in the first, anaerobic compartment in a two-compartment tank, so that a cleaner effluent is finally discharged from the second, aerobic chamber. The inlet compartment of a minimum installation should have a capacity of not less than 500 gallons or a capacity of two-thirds the total tank capacity.

The correct design of a household sewage system includes a distribution box between the septic tank and the disposal field. A distribution box regulates and equalizes the flow in all absorption-field lines. It also serves as a sludge inspection and cleanout manhole.

The soil absorption field, rather than the septic tank itself, is considered the most important element in the success or the failure of this entire disposal system. Where the average rainfall amounts to 30 to 50 inches in a year, the average effluent has been calculated to be as much as 2,000 inches a year. Effluent from a septic tank contains impurities and disease germs, which are converted to harmless matter by air-breathing soil bacteria. However, when effluent is discharged three feet below ground, there is a distinct danger that ground water will be polluted before purification by soil bacteria can take place. ABS (alkyl benzyl sulphonate) water pollution results from the use of detergents containing this chemical.

Soil absorption capacity should first be determined when one selects a disposal field. It is advisable to make a simple percolation test by excavating a hole to the depth of the proposed disposal trenches, about 2 feet deep. Fill the hole with water to a depth of 6 inches and allow the water to seep away. From the average time required for the water to drop 1 inch, one can determine, using the table (see Figure 37.3), the amount of absorption area which is required. In the United States, a daily flow of 100 gallons of sewage for each bedroom is generally estimated.

Septic systems in other parts of the world have been designed to use a small fraction of the amount of water consumed by our systems. Where American toilets use 5 gallons of water in flushing, toilets widely used in other parts of the world can be adequately flushed with 1 quart of water. This is accomplished by installing the septic tank directly below the toilet. This system saves on expensive plumbing pipe and it drastically reduces the size of the septic tank and the absorption field, somewhat minimizing the dangers of water pollution. A few years ago the National Association of Home Builders built a research house in Knoxville, Kentucky, that had a garbage-disposal grinder installed under the toilet. Only 1 gallon of water was required to flush the receptacle, and the plastic sewer drain only needed to be 2 inches in diameter.

Hundreds of houses located in Florida and the Southwest use the sun for heating water. In northern states, solar water heaters are used in summer months and are fitted with auxiliary, wintertime heaters. Once installed, a solar water heater operates every day that the sun shines, practically without maintenance or cost. A solar-water-heating unit consists of an insulated storage tank connected to a solar absorber (collector) plate. As sunshine is absorbed on the black-painted surfaces of the heat-collecting panel, water is heated and flows by natural convection into the storage tank, located above the panel. Water temperature can be raised about 20° by a single passage through the system. Continual circulation is, therefore, necessary before the desired water temperature of 135° is reached. As hot water is less dense than cold water, the storage tank must be at a higher elevation than the heat collector. Thus, cold water flows from the bottom of the storage tank, down through the collector and back up into the top of the storage tank. If the height of the tank above the collector outlet is negligible, reverse circulation may take place at night, with heat being radiated to the night sky.

A solar water-heating installation may best be designed as an integral part of a central utility core. Bathroom and laundry functions can be located on one side of the wet plumbing wall. Cooking and heating facilities may be located on the opposite side. The hot-water-storage tank can be located above the roof, next to the fireplace chimney. Auxiliary water-heating coils may be located in the fireplace and furnace. All of the mechanical controls for these utilities are thus centralized in one convenient core area. The simplest core arrangement where a septic disposal system must be used is illustrated in Figure 37.1. When a compost-privy installation is anticipated, one may consider the design below.

37.4 COMPOST PRIVY-SAUNA DESIGNED & BUILT BY AUTHOR
PHOTOS: EAST VIEW, BOB BROOKS — NORTH VIEW, JOHN RAABE

38 WIRING AND LIGHTING

The journeyman electrician, like the plumber, is a subcontractor whose expensive services can well be dispensed with by an owner-builder. Of all of the building trades, electrical subcontracting represents the greatest mark up of contract price over actual cost. At least $300 can be saved by doing one's own wiring, since the current minimum contract cost for house wiring is 50¢ a square foot or 60¢ an outlet. An electrical contractor is too often trained to think only in terms of outlets and square footage instead of thinking of the efficiency, comfort, health, safety, and the aesthetic effect of his work for a client. Controlling the quality, intensity, and distribution of light in the home is an emotional and personal consideration, best done by the home-owner himself.

According to an article in the magazine *Business Week*, 90 percent of the houses in the U.S. are inadequately wired. This is due mainly to the tremendous increase in the number of appliances in use. It is due also to the obsolescence of older wiring devices and methods. For instance, at least two-thirds of the houses are still wired with a 30-amp service entrance when a 100-amp service entrance is now considered almost minimum.

The traditional fuse-and-switch type of service-entrance panel is no longer deemed satisfactory. Electrical devices and appliances require considerably more current for the split second when they start to operate than when they are operating steadily. A provision for circuit protection should, therefore, include adequate time delay, so that current will not be interrupted upon the accession of electrical loads. Modern automatic circuit breakers employ a combination of thermal and magnetic action. When a predetermined current-time rating is exceeded, bimetal thermal action flexes and trips the breaker. When the current setting is exceeded, an electromagnet breaks the circuit.

One should locate the service entrance panel near the place of greatest load—the kitchen. It is considered good practice to use branch-control centers instead of using the customary method of confining all protective circuit disconnects to one location. With branch-control centers, the individual branch circuits are shorter in length, suffer less voltage drop, and have lower operating expense.

Enough circuits should be provided to carry full power to lights and appliances. The average house requires three different types of branch circuits: general purpose, small appliance, and individual equipment. General-purpose circuits provide fixed lighting outlets and convenience outlets for all rooms except the kitchen and the utility room. At least two 20-amp, general-purpose circuits are required in

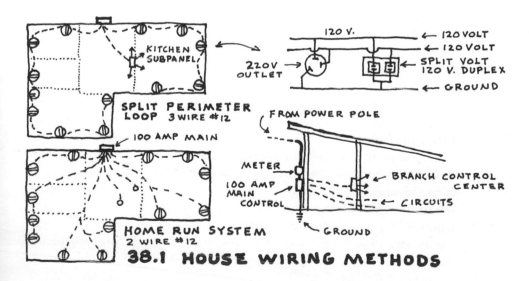

38.1 HOUSE WIRING METHODS

a house. Small-appliance branch circuits serve only kitchen and utility appliances. At least two 20-amp circuits should be provided. Individual branch circuits service the water heater, range, washer-dryer, heating plant, water pump, and so on.

Owing to the expense of copper, English electrical codes specify a type of circuit design known as the split perimeter loop. A 50 percent increase in the amount of conductor which is used will give 3 times the capacity and reduce voltage drop by 75 percent. This type of circuit consists of a 3-wire, No. 12 branch of 220 volts, which is split into two separate, 110-volt circuits at each duplex receptacle. In addition, this 3-wire system provides 220 volts for major appliances at any point on the circuit. One such loop circuit provides lights and outlets in bedrooms; another supplies the living room.

Choosing the proper size of wire conductor is important to prevent dangerously high wire temperatures and a waste of current resulting from a voltage drop. Correct conductor size is best determined from charts that take into consideration the capacity of the load (watts) and the length of the circuit (feet).

Years ago, when electric lights replaced gas lights, wire was merely pulled through the gas pipes, which became electrical conduits, and electric fixtures replaced gas fixtures. Electrical codes still enforce the use of conduits, claiming it is necessary for the prevention of fire and mechanical damage to the wire, but newer, nonmetallic, sheathed cable, called Romex, is far cheaper, easier to install, and entirely satisfactory in respect to its fire and safety provisions.

There are 4 times as many electrical outlets needed at baseboard level than there are needed at ceiling level, yet the customary practice is to install cable in the ceiling. One time-and-material-saving innovation is to predrill a 1-inch hole in the center of all studs, 10 inches from one end, before they are nailed in place. Cables can then be pulled through the studs conveniently at baseboard outlet level.

Some forward-looking designers have come to regard their field as that of light conditioning. Where air conditioners control the temperature, humidity, distribution, and velocity of air, light conditioning determines general light quality, local light quality, brightness contrast, and the glare of light. Unlike the majority of electricians, light conditioners work with illumination rather than with fixture placement. Exact light-source lumens necessary to meet recommended foot-candle intensities can be computed from any lighting table, but maximum lighting is certainly not the sole lighting consideration. Shadow and graduated lighting must also be considered for their aesthetic and emotional effect. The American Public Health Association's Committee on the Hygiene of Housing says:

> Little attention has been given to the effects on the emotions of the lighting of dwellings. The straining "cheeriness" of the professional decorator is probably no more conducive to peace of mind than the amateur's little pools of jaundiced (incandescent) or pallid (fluorescent) subfulgence in a Stygian surrounding. The matter deserves careful investigation, as the results, properly applied, might yield that psychological intangible which turns a dwelling into a home.

From the standpoint of light conditioning, the best form of background or general lighting is that which is indirect. Valances, coves,

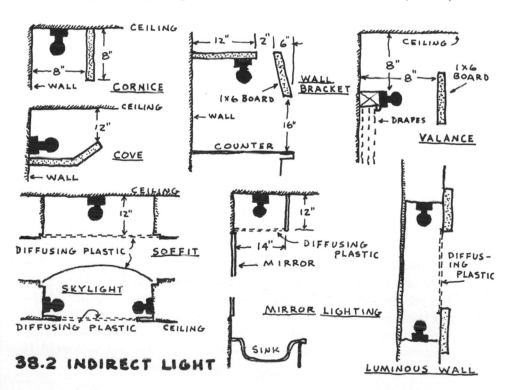

38.2 INDIRECT LIGHT

and cornices are a few of the best means used to redirect room light. The generous source of low brightness found in fluorescent lamps makes them most suited to general lighting requirements. A line source of fluorescent light is much easier to arrange under valances than the point source of an incandescent bulb. Moreover, fluorescent lamps require one-fourth the power that incandescent lamps do to shed the same amount of light. Fluorescent lamps produce far less heat. New "soft white" and "warm white" fluorescent tubes are now available. These newer tubes do not produce the objectionable psychological effects on flesh and fabric tones that earlier types did.

Fluorescent bulbs of the daylight variety, which give off long wave, black light, ultraviolet rays, are a healthier source of light for human vision. All artificial sources of illumination distort the visible light spectrum of normal sunlight and are totally devoid of beneficial, ultraviolet rays. Incandescent light bulbs surpass fluorescent bulbs in this detrimental respect, because they contain a higher red spectrum content. John Nash Ott has written a fascinating book on this subject, entitled *Light and Health,* and he has found through years of extensive private research that the red filter in incandescent light consistently causes plant and animal cell walls to weaken and to finally rupture. This filter blocks beneficial ultraviolet wavelengths, and Ott found that an organism's response to this filtered light was one of altered cell growth, causing either a biochemical or a hormonal deficiency. The condition of *malillumination,* Ott claims, can even be cancer producing.

A matte wall or ceiling finish allows less reflected light (glare) than does an enameled wall or ceiling finish. Eye discomfort and a reduction of visual efficiency are a response to glossy finishes, especially when wall fixtures are used or when ceiling fixtures hang down into the field of vision. Fatigue results when there is a high level of contrast between the directly lighted area, immediately surrounding the task, and the relatively dark surroundings.

Direct lighting is used most often for highly critical lighting tasks, like reading, sewing, food preparation, and so on. Reading and sewing, for instance, require direct lighting in indirectly lit surroundings. The kitchen area should be lit from above, with lamps boxed in over the sink or work counters. Light conditions for table eating should be flexible, capable of following table movements to other positions for other activities.

An endless variety of architectural moods and effects can be created with proper light conditioning. Traditional ceiling illumination has been replaced by a trend toward wall lighting, floor lighting, and spot lighting. Angle lighting is a method of spot lighting which can be used effectively as a concentrated source on vertical surfaces. Deliberate shadows are created with angle lighting, emphasizing textures or revealing architectural details. Spotlight glare is best minimized by deflecting the beam of light from the eyes of the viewer. Spotlights can be reflected off of textured surfaces, such as stone walls, or reflected through planting and foliage to cast shadows on ceilings or walls. When spotlights are placed at the bottom of a pool, the reflected ceiling light is soft and ever-shifting. Outside, spotlighting of a roof overhang seems to extend the inside outward, visually moving the background farther away and making the inside appear larger.

As will be further illustrated in the next chapter, lighting and color offer the owner-builder attractive possibilities for psychological conditioning of one's environment. Rooms can be painted or lit to appear larger, smaller, more intimate, more conducive to conversation, more aesthetically stimulating, more encouraging of work or of study. A familiar instance illustrating this concept would be the 50 foot-candle "mood" lighting of quality, wine-and-dine concept restaurants catering to a lingering clientele, contrasted with that of a 150 foot-candle, direct-lighted chain restaurant catering to a quick turnover of customers.

39 LIGHT AND COLOR

The modern tendency for a professional expert to overemphasize the importance of his particular field of work to the neglect of other equally important fields is as common an occurrence in the building business as it is anywhere else in our overspecialized work world. Illumination experts, for instance, specify an artificial light intensity of 50 to 100 foot-candles for most visual tasks. But experts in the field of light and color conditioning warn against any lighting using more than 30 to 35 foot-candles. They quote ophthalmologists' reports to the effect that visual efficiency rises sharply as light intensity is increased to a level of about 30 foot-candles. Further intensification of light beyond this point of illumination causes visual distraction and glare.

Electrical engineers devise ingenious ways to provide high intensity, daytime, artificial lighting for rooms that are blocked off from natural lighting and which have walls of one tonality (one color value). Millions of dollars are wasted on artificial illumination for want of builders' basic understanding of natural-daylight design. On the other hand, those engineers who attempt to work with natural illumination are often confronted with overcomplicated design formulas from the electrical industry. The complicating factors for the creation of a natural illumination involve light which may be direct from the sun, indirect from the sky, or reflected from the ground. Consequently, the public resorts to more and more complicated and expensive control devices, such as reflectors, glass prisms, plastic-louvered walls, hanging louvers, and glazed diffusing materials, like fiberglass. West Coast nonconformist and expert lighting engineer Foster Sampson tells us that "it really doesn't make much sense to get light through windows in a vertical wall." Skylights and clearstory windows are really more viable alternative-lighting solutions. Skylights, of course, are a very effective means of interior lighting. Improved sky-dome fixtures have recently been developed with two layers of frosted or translucent material, which eliminates bright spots of sunshine and provides air-space insulation.

It is difficult for those of us owner-builders who must use the common glass window to appreciate the many factors that affect harmonious lighting arrangement. Actually, there is only one inexpensive window-lighting-control device available to the building public: that is the Venetian blind. Current lighting research concludes that Venetian blinds are the most effective and flexible means of indoor lighting control. They increase the light level at the far side of a room as much as 34 percent. Ground light and sky illumination are admitted into the room by each reflector-slat. Thus, the Venetian blind is the best stop-gap device employed when there has been a failure or a breakdown of basic lighting design. The owner-builder, without becoming an illumination expert, can achieve comfort and beauty by understanding certain lighting principles. The design criteria for room and window sizes, their placement, roof overhang proportions, and solar orientation must be coordinated with basic, natural lighting conditions, including the latitude and the altitude of the building site and the time of year and the time of day. Later, we will learn how colors also affect lighting design.

A perceptive approach to the lighting problem contrasted with the hit-or-miss approach of the average tract home builder is certainly to be encouraged. One extra day devoted to lighting design relieves one from the need for the services of a competitive electrical subcontractor. An owner-builder might well spread his time for planning and designing his home lighting details over a full season of contemplation. From his building of a scale house model, using a heliodon, the owner-builder can determine his best lighting arrangements. In view of the many years one expects to live in a personally planned dwelling, a day or two devoted to lighting and color design is little enough time to spend.

The first step in considering light and color design is to determine the value of the average outdoor illumination for the specific section of the country in which one lives. The average annual number of clear days gives one some indication of what this value is. Walls, ceilings, and floors receive varying amounts of light. The percentage of light falling on these surfaces which is not absorbed but is reflected should be graded to be somewhere between that of the actual light source and the darker, surrounding surfaces of the room. In order to determine the brightness for a ceiling light, it should be known that the ceiling has a reflection factor of 80 percent. A reflection factor of 25 percent is acceptable on end walls in a room lit with ample daylight. In a deep, poorly lighted room, the wall opposite the window should have a reflection factor of 70 percent. The window wall, as well as its frames and mullions, produces less glare when the reflection factor of surrounding surfaces is high, at 80 percent or more. Floors should have a reflection factor of about 25 percent.

Dark blue and black have low reflection factors. Yellow and white have high factors. The amount of light reflected from various colored surfaces is as follows: white, 80 to 90 percent; pale pastel (yellow, rose), 80 percent; pale pastel (beige, lilac), 70 percent; cool colors (blue, green pastels), 70 to 75 percent; full yellow hue (mustard), 35 percent; medium brown, 25 percent; blue and green, 20 to 30 percent; black, 10 percent.

Colors of short wavelength (green, blue, violet) create the impression of one's feeling cool. Colors of long wavelength (yellow, orange, red) appear warm. Warm colors are sharply focused by the eyes and appear to have qualities of lightness, activity, and advancing movement. Cool colors are less sharply focused and seem to appear heavy, passive, and receding. An obvious but seldom-used rule is to employ warm colors in rooms that are exposed to the north or that receive little sunlight. Cool colors are best used in rooms with a sunlit, southern exposure. Soft, cool colors may best be used in sunny regions, and strong, warm colors may be used in cloudy regions. Rooms receiving a lot of light should be decorated with passive, moderating color. Stronger, brighter colors can be used in small rooms to imply an expansion of the area, whereas low color contrasts and weak light patterns will make large rooms appear less unwieldy.

The story is told of the restaurant entrepreneur who redecorated his dining-room walls from peach to light blue. Soon the employees complained that it was chilly inside the room. The actual temperature, being thermostatically controlled, had not changed. When the walls were again repainted peach and orange slipcovers were put on the chairs, the complaints ceased.

The sustained quest which is required to design a building with balanced light and color is, today, conveniently circumvented by an increasing number of builders who promote the natural-materials approach to decorating and lighting. Wood ceilings are merely stained or varnished. Prefinished plywood wall panels are selected for their wood grain hue. Even imitation wood or cork floor tiles are chosen to match the wood tones that are used everywhere else in the construction. This overbearing use of the natural finish was promoted by Frank Lloyd Wright, whose building interiors had a drab sameness throughout. Wright, who called those who used paint and trim materials "inferior desecrators," clearly overdid his use of wood finishes.

Compare this timid decorative approach to that of a dynamic designer who understands light and color principles, and who uses them to create definitive results. For example, we have a designer-occupant telling us about his choice of light and color for his principal's office at the Waterdloof Primary School, South Africa:

> The character needed for a principal's office is fairly complex. The first impressions of children entering the school are formed here. This demands a friendly, colourful atmosphere. The office is also used for receiving inspectors, teachers, and parents, and should, therefore, be fairly dignified, in keeping with the status of the principal. Office work will require a fairly subdued and quiet atmosphere which is not distracting. Because the room faces into a little court, it may feel rather warm in summer; so a feeling of coolness is desirable. Fairly cool colours are indicated also by the fact that very often only irate parents come to see the principal, and they need to be calmed down.
> The bright, stimulating colours that children like can be used

at a low level, out of the line of vision of the adults. It was decided to use red floor tiles in spite of the fact that red aggravates bad temper. An angry person generally does not look down, whereas a despondent person might.

Psychologists say that 85 percent of our impressions come through our eyes. Light and color, correctly used, will create just about any impression we desire. In a house, we should strive from room to room for a variety and a sequence of impressions, from excitation to sedation. Color should be optically balanced. For instance, a small entry hall, with walls of yellow brick leading to a predominantly blue living room, will complement the cool spaciousness of that room. An excitable impression will be created by a sequence of bright illumination and warm colors, followed by a sudden exposure to cool colors and dim illumination. Finally, a restoration of the bright illumination will create the desired stimulus.

When an impression of sedation is sought, one should decorate the space to be so used with cool colors and with low illumination. Final, gradual restoration of the colors of the active state will give the impression of one having been sedated. A dramatic interior effect can be accomplished by using the maximum degree of color contrast, with abrupt transitions of color value and hue. On the other hand, a static interior effect will be created by using the maximum degree of architectural symmetry and parallelism, of color repetition and continuity.

The degree to which color creates a stimulating or a depressing environment is little appreciated by the average home decorator. Red, for instance, has been found to increase a person's hormonal and sexual activity, as well as to increase restlessness and nervous tension. Time is overestimated by one in red surroundings, and weights seem heavier. Blue, on the other hand, creates the opposite responses. It tends to lower blood pressure and the pulse rate. It is a restful and sedate color. In blue surroundings, time is underestimated, and weights are judged to be lighter. Green seems to subdue nervousness and muscular tension. It is the best color choice for sedentary tasks, for tasks requiring concentration and for meditation. Yellow produces a favorable effect on human metabolism. It is a color sharply focused by the eye, and it is cheerful in appearance. Chrome yellow has been found to be effective for calming shell-shock victims.

This brief discussion of color use neglects to emphasize the wide variety of conditions contributing to one's choice of color for use in one's project. Room color can be chosen on the basis, simply, of one's hair color. A blonde, for instance, looks best against a background of blue or violet-blue, while a brunette looks best with a background of warm, light colors. A person having brown hair looks best amid green surroundings, and a redhead looks best in a room having cool, green-blue hues. A white- or gray-haired person looks best against any brightly colored background.

Color choice can also be made on a psychological basis. An extroverted person, for instance, prefers a high degree of illumination and warm, luminous room colors of yellow, peach, or pink. An introverted person responds to softer, cooler colored surroundings with a lower brightness level. Gray, blue, and green are best suited to this personality type.

Another factor involving one's color choices has to do with the function and the form of a room. It is interesting to note that, psychologically, every color represents tangible, two-dimensional form. Red impresses one as being square in form, yellow as triangular in form, orange as a rectangular shape, green as a hexagonal shape, blue as a circle, and purple as an elliptical shape. The shape of a room or of a building can, therefore, be expressed in color, depending upon whether the room is angular, squarish, or curvilinear.

Room colors should be balanced between warm and cool tones and between active and passive effects. The thoughtful design of a room's form and proportion and the deliberate use of pattern and texture all contribute to color harmony. Primarily, however, color choice depends upon what particular function takes place in the room. A living room, for example, is better decorated in warm tones, stimulating a convivial mood. A more formal atmosphere is commanded by accenting with blue tones. Gray is complementary with all other colors and is used to balance and to harmonize. It is the least distracting of all colors and is most effective for hiding soil and dust. The sense of comfort, warmth, and relaxation so desirable in a dining room can be created by the use of medium-dark, warm colors. Peach is found to be the most appetizing of all tints. Color in the kitchen, which tends to be rather warm during hours for meal preparation, should be light, cool, bright, and generally cheerful, with a semigloss finish. Green and turquoise apparently tend to shorten the passage of time. A restful, relaxing bedroom atmosphere is brought about by the use of cool, light colors. Strong color contrasts encourage early rising. Gray, the peace-maker of colors, is restful when warmed with an admixture of yellow or red. In the master bedroom, a light sky-blue generates a rare atmosphere of allurement, of the illimitable. Lighter, cleaner hues used in bathrooms provoke sensations of vigor and good health. White and blue seem fresh and clean, while pink gives human skin a desirable, luminous glow through reflection. Dark areas or areas for storage should be painted yellow or white for reflective visibility.

Ever since Francis Bacon, inventor of the first color wheel, men have speculated about color harmony and balance. This happens to be a highly subjective field of contemplation, so naturally hundreds of color theories and systems have been proposed. Some basic premises have survived the ages, however, and more recent studies on the subject have produced a simplified charting of color harmony.

In one such simplified charting, there are three dimensions to color: *hue,* referring to the particular, pure color itself; *chroma,* referring to color intensity or saturation—from gray to the pure color (hue); *value,* referring to the light and dark degrees (the tints and shades) of a color. By mixing white or black with another color, a variety of tints and shades are possible.

Faber Birren devised a neat little chart showing the harmonious

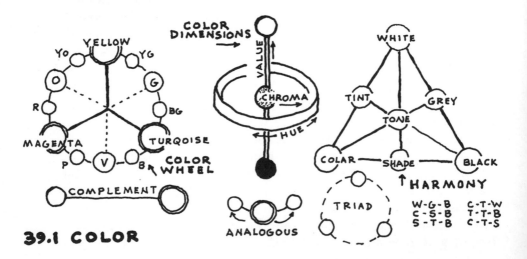

39.1 COLOR

relationships between the seven different color combinations. As noted in the diagram, Figure 39.1, a straight path between any of the seven color items will lead to concordant color harmony.

Basic laws of color harmony require the use of colors in pairs to produce balanced stimulation for color-sensitive eye nerves. Eye fatigue results from overexposure to one color. Relief will result from the introduction of the color's complement or its after image. To explain, eyes fatigued with too much red seek green; when saturated with yellow they pursue purple; when tired of green, they seek violet and red. This complementary relationship forms the basis of our modern color wheel. It is made up of three primary colors (yellow, blue, and red) and three secondary colors (green, violet, and orange). Six intermediate colors, having pure hues, complete the twelve-color wheel.

The complement, which involves a color and its opposite, is the simplest relationship of two-color harmony. Analogous harmony can be prepared by using adjacent colors. Here, interest is created by contrasting values, using light tints and dark shades of one principal color. A third and very commonly used formula for color harmony is the triad. Here, the three colors chosen are equidistant on the color wheel, forming a Y.

A desirable color scheme for any room or area depends much, of course, on its relation to surrounding colors, forms, textures, and patterns. The owner-builder should choose colors with his overall design in mind.

40 DO-IT-YOURSELF PAINTING

If nothing else is learned from studying the forty chapters of this book, it is hoped that the amateur home builder will at least be in a position to scoff at the predominant slogan of organized trades that one should "relax . . . let an expert do it." We should not think of an expert builder as a special kind of person. We should, rather, think of every person as a special kind of builder—planning and working with family and, perhaps, with friends to meet the unique and special housing needs of the growing family. A certain romance is entertained in the home-building endeavor of a congenial and dynamic family.

There are subtle yet vicious aspects to conventional building-construction practices, however—especially since building specialists themselves seldom seem to be aware of the corruptions within their own ranks and of the obsolescences within their own fields. This general observation can best be illustrated, perhaps, with a brief, historical account of the art of painting, as it developed into one of the many building trades.

Credit for being the first painter known to the world will have to be shared by Noah of Biblical fame and Cro-Magnon man, creator of the cave-painted reindeer. After his well-known ark was completed, Noah "pitched it within and without." Pitch, asphaltum, is still used by varnish manufacturers to produce protective coatings. Prehistoric man, it seems, was more interested in the decorative aspects of paint. He mixed simple earth colors with animal fats and painted the walls of his cave home.

Decorative painting assumed an important role in the lives of the Babylonian, Egyptian, Greek, and Roman ruling classes. White lead was formulated and used by Romans as paint as early as 430 B.C. In medieval times, people used paint to decorate and to protect the spokes of their carts and the handles of their various agricultural implements.

With the advent of the Renaissance, guilds were organized with degrees for master, journeyman, and apprentice. The formation of the Old English "Payntour-Steynor" Guild in the thirteenth century laid the ground work for our equally hidebound twentieth-century unions and trade organizations.

Early in the history of painting guilds, items could be found that might well have been written by a contemporary grievance committee:

> In 1488, the Mayor was petitioned to halt the ingress of foreyns (outsiders) from working in the city limits, thereby taking work from members of the guild. . . . In 1502, the Companies of Payntours and Steynors were united into one company. . . . In 1575, the Payntours-Steynors petitioned the Queen against Plasterers who were infringing on their painting work.

In 1581, the Payntours-Steynors received a new charter, seal, and license from Queen Elizabeth. This new ordinance required seven years' apprenticeship, except from those of the gentile class who were permitted to paint for their own private pleasure. No person was allowed to instruct another in the art, unless that person be an apprentice bound for seven years. All work had to be approved by the Masters and Wardens. Masters and Wardens had rights to enter any building construction for inspection and approval. They had the power to impose fines or to destroy the work if it fell below standard. The oaths of all members required them to "keep the secrets of the mistery, and not reveal these same except to apprentices and report all evils to the Company."

In 1606, it was decided that the price for laying color or oil paint upon any flat surface must be sixteen pence per day. Later, the cost was figured by the square yard of the surface to be covered. The work day was stipulated to be from 6 A.M. to 6 P.M. As the Company grew in size and political strength, it started exchanges for the employment of painters. These exchanges became meeting places and finally evolved into our current labor-union hiring halls. Masters and Wardens founded the Institute of British Decorators which, in this country, is called the Painting and Decorating Contractors of America.

Nothing much has really changed in the painting industry from the formation of the first medieval closed shop to the present day. Modern house painters have their own particular brand of price-fixing exclusiveness, and their union, too, is careful to enforce the use of maximum brush widths for specific tasks and to outlaw fast-application, renegade, spray, or roller equipment. Fortunately, one may still work on one's own home. With tremendous advances having been made in the chemical composition of paint in recent years and with unbiased reports coming from countless research agencies, the "secrets of the mistery" are no longer the private property of an inner circle.

The advances in paint manufacture have been very recent. It has not been long at all since a painter, himself, mixed and ground his paints. His first pigment was zinc. Then followed improved titanium dioxide, first used in 1920. From the earliest days, painters used linseed oil as a binder or a vehicle for pigment.

Then, in 1930, what is known in the industry as the alkyd revolution took place. Alkyd resin has all but replaced linseed oil in commercially prepared paints. More recently, the synthetic paint industry has developed other remarkable vehicles, such as phenolics, vinyls, urethanes, silicones, epoxies, acrylics, and latex.

Exterior wood siding should not have to be painted for six years after the first application of paint or stain. Planed lumber requires more frequent treatment than does rough-sawn or rough-planed wood. As a matter of fact, rough-sawn redwood and cedar weather beautifully without any preservative whatever. Experiment-minded home builders have found that any number of natural finishes can be concocted from readily available, inexpensive materials. To create an aged, flat-finish appearance, a mixture of discarded crankcase oil and gasoline has been used with success on rough-sawn siding. A mixture of creosote and pigment stain is another natural finish often used. After several years a second coat of clear creosote and oil mixture is applied to revive protective qualities.

Conventional exterior paint uses titanium dioxide as the white paint solid, and linseed oil and mineral spirits are used as the nonvolatile and the volatile ingredients of the vehicle. But alkyds are more stain, blister, and mildew resistant than is linseed-oil-base paint. The finish coat should have zinc-oxide pigments in it to control the rate of chalking. As a paint ages, it collects dirt, changes color, and chalks. If the paint is correctly proportioned, rain will wash off the dirt along with the chalking. The chalking effect is thus effectively utilized, keeping the paint cleaner and brighter and prolonging its usefulness. Applying a prime coat of shellac or aluminum paint over knotholes and over flat-grain siding effectively seals this material.

Lacquer and varnish films break down sooner with outside exposure than do regular paints incorporating protective pigments. The use of three coats of high-grade spar varnish is a minimum requirement for outside transparent finishes. Butyl phenol resin-based varnishes are superior in hardness, durability, and water resistance to older varnishes.

Polyurethane-based phenolic gum and phenolic tung oil are the best commercially available natural finishes, although they last only two years at the longest. A 1-coat, 8-year, natural finish was developed some years ago by the Forest Products Laboratory which publishes this formula for a 5-gallon batch:

Boiled linseed oil	3 gallons
Penta concentrate	½ gallon
Paraffin wax	½ pound
Color-in-oil	1 quart
Paint thinner	1 gallon

The gallon of paint thinner may be poured into a 5-gallon, open-top can. Put paraffin in a double boiler and heat over water, stirring until melted. Pour this into the paint thinner stirring vigorously. Be sure to keep flame away from the paint thinner. When the solution has cooled to room temperature, add pentachloro-phenol concentrate, then linseed oil. Stir in colors until mixture is uniform, and it is ready for use. For redwood color, use $\frac{1}{2}$ pint of burnt sienna and $\frac{1}{2}$ pint of raw umber, plus 1 pint of pure, red oxide color-in-oil.

Shingle stains can also be used on rough-sawn siding. Linseed oil or oil-modified alkyd coatings are best in California where a porous paint film is necessary. It allows moisture to escape, preventing blistering. Regular shingle stain is composed of a wood preservative, such as creosote, and a color dissolved in solvents containing oil. Preservation of shingles depends upon adequate penetration of the stain rather than the formation of a protective outside film. With regular paint, the adhesion of the film to the surface is more important than the penetration of the wood.

It has been found that 90 percent of all paint-job failures are due to poor lumber with high moisture content. Also, through faulty gutters or faulty flashing, moisture enters the back side of the wood where it condenses, peeling paint and rotting wood. Blistering is common where moisture destroys the paint bond. Paint will adhere best to slow-grown spring wood and will tend to peel when painted on fast-grown summer wood. Paint will also last longer on narrow-band, edge-grain boards than on wide-band, flat-grain boards.

Alligatoring, the final stage of checking, is one common difficulty involving paint mixture. It results from having too much oil in the prime coat. The prime coat should always be harder than the final coat. For this reason, it is important to allow plenty of time between coats, for exposure to the air causes a hardening action (oxidation) to take place. The final coat should be elastic enough to respond to the expansion and the contraction of weather changes.

Lacquer, shellac, spar varnish, linseed oil, or alkyd can all be used on interior wood paneling and plywood. An alkyd is an excellent sealer for plywood. Alkyd gloss or semigloss paint will last as long and retain color as well as any paint now sold. If a colored surface is desired, one of the styrene-butadiene, polyvinyl acetate, or acrylic latex paints is an excellent choice. Already, the greatest volume of water-emulsion paint sold in the United States is of the latex variety. Such paints dry rapidly, are easily applied, have little odor, have good scrub resistance (after hardening), excellent penetration, and good color uniformity.

Latex paints have proven to be the foremost choice for interior as well as exterior masonry surfaces. Styrene-butadiene (rubber latex) was the pioneer among latex paints in this country. It out-performs by far the best oil-based paints and is best used on inside masonry basement walls owing to its water-repellent and alkali-resistant qualities.

Cement-water paint has been the traditionally accepted water-proofing material for masonry surfaces, especially on porous, concrete block walls. The Portland cement content of this material should not be less than 65 percent-by-weight of the total. Fine, sharp silica sand and/or hydrated lime make up the balance. The covering must be applied to damp walls for proper adhesion.

Whitewash is an inexpensive, even older, and still-used masonry coating. The National Lime Association suggests a formula having 5 pounds of casein dissolved in 2 gallons of hot water, 3 pounds trisodium phosphate dissolved in 3 gallons of water, and 3 pints of formaldehyde mixed in 3 gallons of water—all added to 8 gallons of lime paste, which is 50 pounds of hydrated lime mixed with 6 gallons of water. The lime coating is applied to damp walls and dries to an opaque, hard, dust-free finish.

Polyvinyl acetates and acrylics have excellent color-retention and water-resistant qualities on outside masonry surfaces, on asbestos cement, and on asphalt shingles. Resin-based urethane is a new concrete floor finish that is claimed to outlast other floor finishes 3 to 5. Chlorinated rubber and phenolic are used where the floor is exposed to acids, alkalis, salt, or other corrosive elements.

There are a number of factors that influence the proper choice of masonry paints. Cement-water paints, for example, are more suitable for damp, new walls containing open-textured surfaces and water-soluble alkaline salts. Resin-emulsion paints are better for dry, close-textured surfaces, such as cast concrete, asbestos-cement siding, tile, and so forth. Oil paints are best used on masonry surfaces that are dry at the time of painting and which remain dry afterward.

Corrosion (rust) is a common problem when using exposed metal in building construction. Zinc-dust paint is about the only primer that will adhere satisfactorily to galvanized metal surfaces. To prime steel, red lead (iron oxide), or zinc chromate in linseed oil, alkyd or phenolic vehicles are satisfactory. A linseed oil or alkyd finish coat is then applied. The best preservative for tarpaper roofs is a commercial asphalt varnish mixed with aluminum powder. Upon application, the aluminum flakes float to the surface, creating a metallic finish. These paints combine durability with reflective roof-coating features.

These recommendations on paint materials treat only one small segment of the total painting picture. Equally important, a painter must understand the paintable characteristics of various surfaces. One must choose and use correctly the different tools of the trade. Finally, the painting procedure must be correctly established, such as painting the house exterior before the interior is done, and painting the ceiling before the walls before the woodwork before the floor before the baseboard.

To insure proper paint absorption, a good quality paint brush will have more long-length bristles than it will have short-length bristles. Stiff and soft bristles are correctly proportioned in a brush to allow for proper paint retention. Some of the newer quality nylon brushes are as good as the traditionally superior Chinese hog-bristle brushes.

In many cases, a paint roller will do a faster and a smoother appearing job than a brush. It can be used for applying any kind of paint, but the surface to be painted will determine the roller cover nap length; the smoother the surface, the shorter the nap should be.

CONCLUSION

In view of the multitude of technical details covered in these forty chapters, the would-be owner-builder may be inclined to throw up his hands and to declare that this is all too much for a realistic response. But consider, first, that the details set down herein are meant to meet the needs and interests of a large and varied owner-builder audience. All of the details suggested in this book are not meant for each and every prospective builder. Each owner-builder will, rather, select from this material only that which will help with one's proposed building project. Perhaps one will utilize but one-tenth or one-twentieth of the suggestions offered in these chapters. A student in a university architecture school is required to become familiar with all of the use-and-beauty building ideas which have evolved in every country from the year one until the present, but this is not for the average owner-builder who will, most likely, build just one home for one family.

Building a home should not be the formidable project that it seems to be for some owner-builders today. Ideally, land should be assigned without charge by one's local community for house building and for agriculture. Money should be loaned by a community bank without

349

excessive charge. There should be counsel and friendly help from neighbors who have already built their homes. For those reading this who may think this sounds foreign or even subversive, just remember that early American pioneers won their frontiers and built the West in very much this same way.

But, of course, this is not the situation in most parts of this country today, or in most parts of the world, for that matter. Most would-be owner-builders find themselves hampered at the outset by not being able to secure land, the necessary capital, or the requisite expertise.

The problems associated with access to land were clearly enumerated to my freshman class of architecture at the University of Oregon by Dean Willcox. Walter Ross Banner Willcox believed, as did the economist Henry George, that the land problem was very much linked with unjust taxation and with the rampant land speculation that ensued from the excessive levying of taxes. In the preface of his little book *Taxation Turmoil,* Dean Willcox wrote:

> The following pages were written by an architect. They were written in a spirit of protest against what seems to be a settled policy of those who direct and influence the affairs of government. The policy referred to is that of ignoring the benefits which might come to this country and the world from a thorough and sincere questioning of the efficiency of Taxation as a means of securing public revenue.

Thirty years have passed since Dean Willcox spoke to Oregon's class of fledgling architects. One wonders how many of those students, then or later, carried this spirit of protest. In his own life, Dean Willcox not only reflected respect for civil disobedience—an important trait for young idealists—but he exemplified allegiance to other vital aspects of life as well. For instance, the Willcox home on Kincaid Street in Eugene had an exceptionally well executed, owner-built interior; outside, however, for the benefit of the property tax assessor and the local building inspector, his Victorian mansion was shabby and unkempt.

As we students became aware in our studies, it was a totally new concept to many of us that the County Building Department and the Tax Assessor's Office might effectively—if not actually—be in collusion. According to the Uniform Building Code, a building permit is required if the extent of remodeling or repairing a building exceeds $20. The requisite permit is then processed by the tax assessor and any improvements thereby increase one's tax rate accordingly.

Our whole system of confiscatory taxation horrified Dean Willcox. He, and others like Henry George, believed that private ownership and speculation in land were the primary evils of western civilization. He believed that instead of property improvement tax and income tax, there should be a single revenue, called land rent, to operate community services.

The tax assessor and the building inspector, instead, come down hard on the defenseless poor man. His quest to replace the dead-end, rent-paying syndrome with an owner-built home can be totally squelched at the local department of health and safety. On a national average, compliance with the Uniform Building Code increases a dwelling's cost by $1,000. In numerous instances, this sum represents for the owner-builder the total financial reserves for the family—not a surplus to be wasted on antiquated and unrealistic code requirements. The $1,000 owner-built home may be labeled "substandard," but it most likely proves to be for that owner-builder a more satisfying housing solution than paying rent for a so-called legal and approved house. In a very real sense, the Building Code perpetuates landlordism, especially in states like California.

In 1964, the California State Department required every county in the state to adopt the Uniform Building Code. Before that time, poor counties offered code exemption in agricultural or sparsely settled areas. In the author's county of residence, the County of Madera, for instance, no code compliance (prior to state regulation) was required of new housing built on five or more acres if the building was located over 50 feet from the nearest property lines.

The building law was originally introduced primarily for tax-gathering purposes, but the influence that other private groups exert in this ripoff drama cannot be discounted, either. The Realty Board exerts tremendous political influence in Sacramento. It is to their commission-collecting advantage to maintain high property values. Middle-class standards will be preserved—and reserved—democratically for those who can afford it. The poor, those of racial minorities, the hippie, the nonconformist, and a great many wage earners must all be content with an urban or substandard rural rental.

Building-contractors associations have much to gain from hard-line code enforcement. After all, they helped to write the code, and the code-enforcing inspector is astonishingly often a one-time, bankrupt, building contractor. The general building contractor and the city hall building department share a mutually beneficial, tacit agreement to manipulate bureaucratic pressure, to increase profits, and to preserve the status quo. An innovative, maverick building contractor must be closely watched, for he is as much disliked by those behind the desk as is the troublemaker owner-builder.

Construction materials manufacturers have a powerful influence on building-code requirements. In 1940, a lobby for the National Lumberman's Association in Washington was responsible for stopping the research and the construction of rammed-earth government buildings in the Southwest. As there was no Washington-centered pressure group organized around the sale and distribution of earth, work was not only discontinued on these government buildings but earth construction per se was virtually outlawed for all building done with this universally available material. One can build an earth-wall home in California —but at a cost several times greater than standard frame construction due to entirely unrealistic amounts of steel reinforcement which are required by the building code. Only more wealthy home builders can afford to build with the least expensive of all building materials —earth.

Adoption of the Uniform Building Code in 1964 set up a gravy train that seats many. Some of the riders are strange bedfellows, all intent to get in on part of the action. Imagine architects, engineers, and building designers, again, effectively if not actually conspiring together: a group of plan preparers. Architects and engineers are, of course, members of powerful associations working for their own special interest. In the high desert country of Kern County, California, for instance, all construction plans (even those for a minimal, single-family dwelling) must be prepared by a registered architect or engineer. Building designers are a somewhat more pitiful lot, but deserve dubious recognition for a valiant lobby struggle to gain status and power. Their association is called the American Institute of Building Designers. As either drop-out architects or glorified draftsmen, they attempt to identify, at least in name, their "American Institute" with the more prestigious American Institute of Architects. The A.I.B.D. fellows feel that they alone, in league, of course, with big brother architects and engineers, should be allowed to prepare working drawings for new construction or for structures to be remodeled. At this writing, they are not yet accredited by the State of California, and their Institute is nothing more than a fee-setting, mutual-aid society.

Powerful labor unions are not without their representation on the Board of Building Officials. Few leniencies are permitted the owner-builder who wishes to plumb or to wire his own house. Code enforcement is stringent, because the trades themselves are antiquated. Enforcement is a last-ditch effort for building-trades protection. A twelve-year-old child can plumb the average house using modern plastic pipe, plastic fittings, and bonding adhesive.

Finally, the control of money resources has become a decisive factor in the erection of shelter. Banks loan money only on code-approved buildings. Payments are progressively made to builders as various stages of the construction are reached and are accepted by local building inspectors. Banks seldom loan to an owner-builder. They prefer to work with a licensed building contractor who is willing to post bond for cost and building completion.

Someone recently figured that a $25,000 tract house, bought through a bank at an average 1974 rate of 9 percent interest will, at the end of the 25-year mortgage period, cost in excess of $100,000. The interest alone comes to more than the original cost of the house. Add to this the required homeowner's insurance premiums, property taxes, and home-maintenance costs, and one finds that, over the quarter of a century of these payments, the original price tag for his house has quadrupled!

All of the above special-interest services are expensive and absolutely unnecessary to the house-needy family. However, the Building Department, in effect, sets up the unwary home builder like a brace of clay pigeons and sanctions the legalized pot-shots that pursue his efforts. Anyone who has built a house in California in accordance with the building ordinances knows of this. The building code, with all that it represents and all that it protects, is an outrage. People should rally around its reform with the same fervor that they support peoples' initiatives. A peoples' lobby or a consumer protection agency needs to be formed to avert the massive building ripoff that currently takes place when a poor man stands up for his right to supply his own housing needs.

Currently, some owner-builders have organized to better deal with the problems and the limitations represented by the Uniform Building Code. In Mendocino County, California, the "United Stand" group seeks such code reform, specifically allowing the individual property owner the freedom to build however he wishes, so long as he remains the sole occupant. When the building is sold, traded, or rented, it must first be brought up to code standards. United Stand also proposes that alternative methods of waste disposal be permitted so long as public health is not adversely affected.

Limited, expensive land and building-code restrictions are only two factors that hamper or outlaw owner-builder projects in settled areas. An urban, two-day-a-week, two-weeks-a-year home-building program is next to worthless. One needs a block of free time to build a house. One also needs the energy and the well-being that can come only with good nutrition, fresh air, clean water, and sound rest. These natural prerequisites form the backbone of the alternative, rural-living solution to peoples' housing needs. Very soon in the construction process an owner-builder finds that positive resources are required that come only from a more or less natural and friendly community in a rural environment. The two most important of these resources are freedom and health.

A rural setting can provide an owner-builder with a set of dovetailing, supportive circumstances. A family may buy acreage in the hinterland where land is not so expensive. Taxes will, therefore, not be so high. Building regulations may, hopefully, be almost nonexistent. Therefore, only moderate construction funds need to be amassed. The land may be made to be productive with the intention that less cash will need to be earned for foodstuffs. More time and money may, consequently, be spent on building and on land development. Nutritious food raised on the land will improve a family's health immeasurably and more energy will be generated for greater homestead development.

In a few years a family should be happily situated on its own land, in its own debt-free home. How it should go about developing the land—the garden, orchard, pasture, woodlot, water supply, fish-culture ponds, fencing, barn, shop, and outbuildings—has been the subject of my second, 200-page book, *The Owner-Built Homestead,* published in 1974 by Owner-Builder Publications.

The idea of a family earning its economic necessities from a homestead, with a part-time money income to supply amenities which cannot be family-produced, goes back to the Depression years when President Roosevelt's Federal Security Administration dabbled in subsistence farmsteads. But a much more significant contribution to this back-to-the-land movement was made by pioneers like Ralph Borsodi and Milton Wend.

Ralph Borsodi and the books he wrote in the thirties and those he has written since have helped to shape the homestead movement. Economist Borsodi established his family homestead 25 miles above New York City in 1921, for he saw the need for small-scale technology to attempt a revival of productive living. In 1929, he wrote his famous critique of modern culture, *This Ugly Civilization,* and in it he suggested that the small, self-subsistent homestead was a human and constructive way out of the urban nightmare that was then developing. All of this was popularized in his *Flight from the City,* in 1938.

Borsodi, in effect, cast the idea and the reality of the modern homestead into the social pool of the 1930s. The ripples of that act continue to widen. Some of those affected have been Milton Wend, Ed Robinson, J. I. Rodale, and Mildred Loomis. Borsodi established the first School of Living near Suffern, New York, in 1937, to do research on how to live a personally enriched life, how to build homesteading communities, and how to develop a curriculum for a new education for living.

Milton Wend, now of Edgartown, Massachusetts, was a trustee of the first School of Living. His experiences and ideas were reported in his fine book, *How to Live in the Country Without Farming.* The book was then widely read and remarkably influential in directing many World War II veterans to return to the land. This book is today, fortunately, being republished.

Ed Robinson grasped a few of these ideas from a School of Living brochure entitled, "Have More Vegetables," and developed his famous Have More Plan and his country bookstore. After a flourishing business, these works were discontinued in the 1950s.

J. I. Rodale visited the School of Living in 1938 and there saw the composted gardens, the use of whole foods, the grinding of grain into flour and cereals, and the regular baking of whole-meal breads. He went back to Emmaus, Pennsylvania, and later changed his publishing emphasis to gardening and homesteading. The impressive growth and influence of the Rodale Publishing enterprise is well-known today.

Mildred Jensen Loomis was assistant educational director of the Suffern School of Living, 1938 to 1940, and later continued that work avocationally at her home at Lane's End Homestead, near Brookville, Ohio. Her editing of journals (*The Interpreter* and *Balanced Living*) began in 1944 and continues in 1974 with *The Green Revolution*. The numbers of people who have been influenced to take up the homestead way of living from these publications and from Mrs. Loomis's edited book, *Go Ahead and Live!*, are uncounted.

During the Depression of the 1930s and 1940s, books like Kains's *Five Acres and Independence* carried on Borsodi's early vision. Unfortunately, these early writers and promoters of the country life did not produce a dominant trend in our country. Why? The reasons for this may be many. The technological drift of our modern day had attained a momentum which could not be stopped by a trickle of counterculture ideas. And the form and content of the discourse about rural living in the thirties and forties were of a pre-Depression vintage. Traditional living patterns were merely dressed up in a country-living format and were presented as a bona-fide original. Many would-be homesteaders were turned away or became disillusioned. There was also no qualified professional or educational assistance in the homestead movement. One exception to this condition was a mere architectural competition for a productive homestead, sponsored by the *Free America* magazine.

So, the first wave of homesteading interest in the late thirties and forties subsided. Some of the leaders seemed to drift into singular, specialized aspects of the movement, such as organic gardening, nutrition, or craft production. This was probably abetted by the seemingly narrow and limited understanding of homesteading as being other than a whole way of life. People thought that no earth-shaking revolution or revelation could ever come out of a potato patch. Moreover, high employment and the Social Security benefits offered by the Great Society of the 1950s and 1960s with a war-making power elite running its government and its institutions actively dissuaded people from a life on the land. Government handouts in the city were easier to accept than was living in the cracks with one's own wits, using the rural margins of a spoiled, affluent society.

Today, however, the war-rampaging, raw-material-devouring, land-procuring, money-controlling, power-hungry elite are due to be unseated, as catalogued and revealed in all of our worldwide communications resources. The overpopulated sinks of the cities' poor naturally spawn riots and a clamor for change. There must be some shift toward equilibrium. The homestead and the village community life that has been repeatedly disrupted by predatory human ego since early in the history of our civilization and which has virtually been threatened with extinction by the Industrial Revolution must, in measure, be restored to the world's billions who know nothing of elitism and much about poverty, suffering, and starvation. Along with our present worldwide communication, our potential for free, peaceful enterprise, our scientific vision, and our ingenious technology, we can—if we will—build a more livable world for all that belongs to life on earth.

GREENBAUM HOMESTEAD IN MT. ANGEL, OREGON
PHOTO BY JOHN RAABE

PHOTO: JOHN RAABE

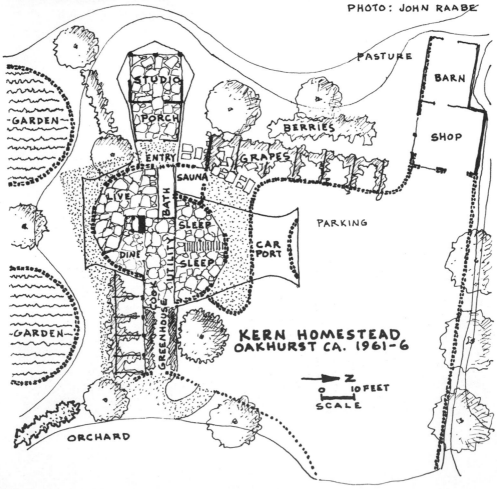

PASTURE

BARN

SHOP

GARDEN

STUDIO

PORCH

BERRIES

ENTRY

GRAPES

SAUNA

LIVE

BATH

SLEEP

PARKING

DINE

SLEEP

CAR PORT

GREENHOUSE

UTILITY

COOK

GARDEN

KERN HOMESTEAD
OAKHURST CA. 1961-6

N

0   10 FEET
SCALE

ORCHARD

ABOUT BOOKS

According to current Bureau of Census statistics, 160,000 American families build their own homes each year, and an estimated one-third of the world's people house themselves in constructions built with their own hands. Lacking professional construction assistance or governmental subsidy, this multitude of worldwide owner-builders must somehow gain access to viable building information. Rural, indigenous, third-world peasant communities do have access to building tradition and knowledgeable assistance cooperatively shared among themselves, while people of the industrialized, free-enterprise nations are less fortunate. They must somehow sift through the heap of misinformation and the propaganda of commercial self-interest to arrive at a proper building plan and building program.

Most Western owner-builders depend upon materials suppliers and popular "home" magazines as sources for building expertise. *House and Home* and *Sunset* may illustrate more examples of contemporary, classy housing and interiors than *American Home* or *House and Garden*, but virtually all of these building magazines are unfortunate sources for the owner-builder seeking information for a site- and climate-attuned home designed for specific, individual space requirements. The best English-language building design periodical, *Architectural Design*, is published in England at 26 Bloomsbury Way, London, WCIA 255.

Books, of course, have become Western man's traditional and commonly accepted source of information. Books can be of tremendous assistance to the owner-builder, or they can become a costly and misleading burden to the unwary. Some books excite and stimulate owner-builder effort, while others totally discourage self-actualization. A proper bibliography should, then, necessarily include titles for readers to avoid, as well as titles for recommended reading.

As this book is being compiled (spring 1975), there exists a daily influx of new building literature destined for this year's 160,000 potential owner-builders. The do-it-yourself book market has reached explosive dimensions. Anyone who randomly selects even a small part of this literature for his working owner-builder library ultimately runs a real risk of confusing theory with practice, means with ends, and the word with the deed. Like the indiscriminate accumulation of building tools, the accumulation of building books can become a vicarious substitution for doing the building itself. For this reason, the author suggests a select reading list for those wishing to explore more thoroughly a particular aspect of owner-building. A general listing is included under each section of chapter groupings.

Freedom to Build is a good beginning book with which to grasp what self-actualizing housing is all about. It is edited by John F. C. Turner and Robert Fichter and is published by Macmillan (1972). This book speaks of the economic plight of self-help housing in both third-world countries and in our own contrastingly affluent but nonetheless house-needy countries. It tempts one to read further into the subject of indigenous building with books such as *Shelter and Society* by Paul Oliver (Praeger, 1969); *Architecture without Architects* by Bernard Rudofsky (Museum of Modern Art, 1964); and *Craftsmen of Necessity* by Christopher Williams (Random House, 1974). The latter is an exceptionally well-prepared, contemporary documentary on the spirit of building in primitive cultures. A best-seller American counterpart of the Williams book, *Handmade Houses* by Art Boericke and Barry Shapiro (Scrimshaw Press, 1973) $13, falls far short in its presentation of would-be vernacular (a better term is funky) hip culture architecture. Must we conclude from presentations such as this that contemporary Americans can only achieve the casual, "natural" effect in their building with highly contrived, skillfully executed, artsy-craftsy effort? Can reader interest only be aroused with presentations which are printed on slick, full-color pages that are outrageously priced? At best, such books are interesting coffee table conversation pieces, but they are unsuitable for serious owner-building.

Two excellent books on primitive building should be included in this section: Amos Rappoport's *House, Form and Culture* (Prentice-Hall, 1969), and *Village in the Sun*. With clear drawings and precise text, Rappoport masterfully traces building forms to their modern counterpart.

For a more general treatment of the historical aspects of building, James Fitch's *American Building and the Environmental Forces That Shape It*, published by Houghton-Mifflin (1972), cannot be too highly recommended. Other all-time favorites of this author which are recommended for reading: A. J. Downing, *The Architecture of Country Houses* (Dover, 1969, originally published in 1850); James Kenward, *The Roof Tree* (Oxford, 1938); D. E. Harding, *The Hierarchy of Heaven and Earth* (Faber and Faber); and Henry Bailey Stevens, *The Recovery of Culture*. An owner-builder can learn much from various expositions on historical building practices. For example, I had occasion to trace the development of cast rubble wall construction in preparation for chaper twenty on stone walls. Back in the 1840s, a Mr. Goodrich devised a system of movable wooden wall forms into which he packed a sand, gravel, lime, and stone mixture. In 1848, Orson Fowler elaborated on Goodrich's wall-building method and wrote about it in *The Octagon House*, reprinted by Dover in 1973 ($3).

Then, in 1921, New York architect Ernest Flagg added an improved external bracing system to the wall form and wrote about "his" system in *Small Houses, Their Economic Design and Construction* (1932). Soon after Flagg's work was published, Frasier Peters came out with his *Houses of Stone* and, later in 1949, with his *Pour Yourself a House*, in which he elaborated his own variation of stone wall building. Architect Frank Lloyd Wright copied Goodrich's original forming system when he built his Taliesin West design headquarters in Arizona and wrote about stone building in *The Natural House*. Scott and Helen Nearing substituted lighter-weight plywood for the planking formerly used and described their stone wall building experiences in their animated account of *Living the Good Life* (Social Science Institute, Harborside, ME). Another husband-and-wife team, Lewis and Sharon Watson of Sweet, ID, have written (1974) of their poured rubble house in *Our House of Stone*. The history of form-cast stone walls makes interesting reading, especially for those owner-builders contemplating the use of this type of construction.

The house model shown in this bibliography was designed by the author as a prototype, visually illustrating the maximum number of planning, design, and construction features mentioned in this text. The basic floor plan is shown in Figure 8.6, and the cross section is shown in Figure 2.1. The first nine chapters of *The Owner-Built Home* are concerned with the physical characteristics of one's building site and the climate components of one's building environment. The photographs of the model above attempt to articulate such pertinent considerations as (1) planted windbreaks along cold north exposures, (2) earth berm insulation of the building's north wall, (3) roof insulation accomplished by use of sod covering, (4) natural ventilation promoted by proper roof slope and by appropriate window openings and vents, (5) south-facing building exposure, maximizing interior receptivity to winter sun while minimizing interior heating by summer sun, (6) dark-colored slate floors for wintertime accumulation of solar heat.

Also pertinent to the aspects of building site and climate control in the first section of this book are the following, listed in the order of their importance:

Robinette, G. O., *Plants, People and Environmental Quality*, U. S. Dept. of Interior, 1972

Givonni, *Man, Climate and Architecture*

Roger, T. S., *Design of Insulated Buildings for Various Climates*

Olgyay, A. and V., *Design with Climate*, Princeton, 1963

Harada, Jiro, *Japanese House and Garden*

Ott, John, *Health and Light*, Devin Adair, 1973

Neutra, Richard, *Mystery and Realities of the Site*, 1951

Van Dresser, Peter, *Landscape for Humans*, Biotechnic Press, 1973

Simons, *Landscape Architecture*

Eckbo, Garnett, *Landscape for Living*, F. W. Dodge, 1961

Costing, H. J., *Plant Communities*

MacKaye, Benton, *The New Exploration*

Aronin, J. E., *Climate and Architecture*, Reinhold, 1953

Colvin, B., *Land and Landscape*

Baker, J. H., *Windows*

Sheppard, Richard, *Building for Daylight*

Newbury, L. H., *Physiology of Heat Regulation*

Putnam, J. Pickering, *The Open Fireplace in All Ages*, 1886

Graham, E. H., *Natural Principles of Land Use*

Geiger, Rudolf, *Climate near the Ground*

Conklin, G., *The Weather-Conditioned House*

Pierce, Josephine, *Fire on the Hearth*

Besides the books listed above, many of which are now out of print and available only through public libraries, a number of excellent research bulletins have been issued by government and university research stations. The results of *The House Beautiful Climate Control Project*, which took place from October 1949 through January 1951, were issued as a bulletin of the American Institute of Architects, March 1950, and this work on climate is probably the most valuable of its kind for owner-builders. Unfortunately, the bulletin is now out of print and only available as original articles in *House Beautiful* magazine. Following is a list of public agencies who have contributed much on site and climate research, again listed in the order of their importance:

National Building Research Institute, Pretoria, South Africa—building climatology
Texas A & M College—ventilation
National Research Council, Building Research Advisory Board, Washington—light and shade

Housing and Home Finance Agency, Washington, D.C.—summer cooling
Building Research Station, Garston, Herts, England—fireplace heat
University of California, Davis—summer cooling
University of Illinois, Small Homes Council—central heat
Stanford Research Institute, Palo Alto, CA, and the University of Arizona—space heat
Commonwealth Experimental Station, Australia—building climatology
American Society of Heating and Ventilating Engineers, NY—summer cooling
Building Research Board, London—light and shade
Clay Products Association, Austin, TX—summer cooling
United States Public Health Service, Washington—light and shade
University of California, Davis—planting design

Chapters ten through sixteen of *The Owner-Built Home* concern the design and the function of a house. Included in this section is a general discussion of the psychology of space. The author's house model, illustrated herein, incorporates many important design and function concepts. Notice, for instance, the free interior circulation, the open play of various areas. Living space revolves around a central heat-and-cooking core. Sleeping is tucked into an upper-level loft. The utility functions of bathing, sauna-bathing, clothes washing, and toileting are all located in a three-level, towerlike structure that includes a ground-level composting chamber and an encompassing sun-pit greenhouse. Represented for the inspiration they contributed to this house design are such books, listed in the order of their importance, as:

Sommer, *Personal Space*
Hall, Edward, *The Hidden Dimension*, 1966
Hall, Edward, *The Silent Language*, Fawcett, 1965
Mollar, *Architectureal Environment and Our Mental Health*, 1966, and *Sociology, Building and People*
Pye, *The Nature of Design*
Ricci, Leonardo, *Anonymous (20th Century)*
Grillo, *Design*
Rudofsky, Bernard, *Looking Through the Picture Window*
Kennedy, Robert Woods, *The House*
Massee, William, *The Art of Comfort*, 1962
Moholy-Nagy, *Vision in Motion*
Goodman, Paul, *Communitas*
Pearce and Crocker, *The Peckham Experiment*, 1943
Venturi, Robert, *Complexity and Contradiction in Architecture*
Scott, G., *The Architecture of Humanism*, 1914
Zevi, Bruno, *Architecture as Space*
Sweeney, *Antonio Gaudi*, 1960
Steiner, Rudolf, *Ways to a New Style in Architecture*, 1927
Gutheim, *Houses for Family Living*
Beyer, Glenn, *The Cornell Kitchen*, 1952

The third section of *The Owner-Built Home,* chapters seventeen through thirty, covers the subjects of materials and skills. This section, as the author's house model illustrates, employs such important structural features as the stone-faced, concrete slip-form wall, which is curved to better resist the hydrostatic forces inherent in earth-banked walls; and the light-duty masonry floor, designed to promote no-draft air circulation through its continuous, sub-floor air space. Simplified roof and wall framing methods are included in this design to facilitate fabrication by unskilled owner-builders.

Few good books applicable to owner-erected construction are available for this section of our discussion. Mostly, the subject relative to building skills is either padded with unusable, antiquated building methods, such as those found in the Audel manuals, or contains reactionary building ideas which, however novel, fail to prove satisfactory with experience and the passage of time. A best seller, Rex Roberts's *Your Engineered House,* falls into this latter category. Innocent, unsuspecting home builders soon find themselves in grave economic and technical difficulty when they attempt to erect a structural frame to accommodate the insulation of their building using Roberts's suggestions. Burroughs once remarked, "To treat your facts with imagination is one thing, but to imagine your facts is another!"

Possibly the finest work available on materials and skills is the two-volume *Fundamentals of Carpentry* by Walter Durbahn. Published by the American Technical Society, it has survived twenty-three printings since 1947. Other recommended publications of reference in this section are, in the order of their importance to the owner-builder:

Kahn, Lloyd, *Shelter*, Random House, 1973
Central Building Research Institute, New Delhi, India—roofs and floors
Virginia Polytechnic Institute—wood fastening
Middleton, *Earth Wall Construction*, Sydney, Australia, 1952
United Nations Housing and Town and Country Planning—rammed earth
Low-Cost Wood Houses, U.S. Department of Agriculture, Forest Service, 1969

American Wood-Preservers Institute, Washington, D.C.
Department of Agricultural Engineering, Brookings, SD—earth walls
Portland Cement Association—tilt-up and sliding molds
Lumber Dealers Research Council, Washington, D.C.—framing
Structural Potential of Foam Plastic for Housing in Under-developed Areas, University of Michigan, 1966
National Lumber Manufacturers Assn., Washington, D.C.

The final section, chapters thirty-one through forty of *The Owner-Built Home*, has reference to a building's form and structure. Again we look to our model. Structural detailing is not possible in model form. One can only imagine, after viewing the general plan of this house, how structural foundation, floor or wall components will look, or how plumbing and electrical systems are to be assembled. Reading for one such system should include Alexander Kira's *The Bathroom*. This book exemplifies the very best research techniques and thinking available on this one specific segment of the house. McGill University's booklet *Stop the Five Gallon Flush!* is a comparable study on the same subject. Also included for reading in this section are:

Excreta Disposal, World Health Organization—composting
Soil Construction, S. Cytryn, State of Israel, Ministry of Labor, 1957
Grade-Beam and Pier Foundations, University of Pennsylvania
Birren, Faber, *Color, Form and Space*
Construction and Equipment of the Home, American Public Health Assn., 1951
Field Applied Paints and Coatings, National Research Council, 1959
Wood Floors for Dwellings, Forest Products Laboratory, Madison, WI, 1961

Other books by the author, available from Owner-Builder Publications, Box 550, Oakhurst, CA 93644, include:

The Owner-built Homestead, 1972 ($5)
The Code: Politics of Building Your Home, 1975 ($5)

INDEX

Adobe: block, 141–49; block, molds, 148; block, production, 148; block, testing, 146; dome, 113; mixers, 148–49; mortar, 149; stabilizer, 143; strength tests, 142
Air circulation. *See* Ventilation
Air conditioning: fans, 40; and health, 34; natural, 22, 23; and solar orientation, 37; subterranean-tempered, 22; units, modern, 34–36. *See also* Cooling
Alcoves, 125
Asphalt: emulsified, 247, 258; floors, 280–81
Attic(s): fans, 40–41; ventilation, 30; vents, 30

Back-fill building, 112
Baer, Steve, 259
Balloon-frame construction, 222
Bamboo, 7, 8; as reinforcement, 244–45
Basement(s), 270–71
Bathing, 322–24
Bathroom(s), 130, 318–21; units, plastic, 252
Bays, Jack, 243–44
Beam(s), 222; box, 300; cantilevered, 230; design, 299
Bed(s): compartments, 129; room size, 129; and sleeping, 129
Billig, Kurt, 6, 241–42, 282
Binuclear house, 13–14
Birren, Faber, 340–41
Bitumul, 143, 144, 244
Bloc, Andre, 106
Block(s): adobe, 141–49; crushing strength, 147; details, 184; header, 183; laying, 154–55, 182–84; lightweight, 186; masonry, 179–91; masonry, cleanup, 191; masonry, designs, 188; masonry, insulated, 187–88; masonry, laying, 182–84; masonry, moisture penetration, 188–89; masonry, reinforced, 189–91; masonry, surface bonding, 185–86; masonry, walls, 29; masonry, wall separation, 189; mortarless, 184; patterned, 186–87; pressed, 150–56; presses, 150–52; rice hulls and cement, 239; size, 185; stabilizing, 144–45; sulfur, 8, 202; tests, 146; walls, building, 183–84. *See also specific blocks*
Boester, Carl, 273
Bottle technique of construction, 246
Box beams, 300

Bracing tests, wall, 290
Brick: details, 190; laying, 182; masonry, 179–91; masonry, reinforced, 189–91; and mortar joints, 181; and reinforcing bars, 190
Brooks, Bob, vi, ix
Building: climatology, 18–24; codes, 3; construction, 4, 5; labor, 3; loans, 2. *See also* Climatology
Building materials. *See* Composite materials; Salvage materials; *specific materials*
Building site, 4, 11–17; analysis, 12; evaluation sketch, 12; and house planning, 13; physiognomy, 16; and pole-frame construction, 230; survey, 11; and view, 15
Burlap sacks, as reinforcement, 241

Candela, Felix, 193
Cantilevered: beams, 230; decks, 230; foundation systems, 276; room extensions, 230
Capillarity: in floor-slab construction, 281–83; in walls, 291–92
Carter, Lawrence, 111–13
Caudill, William, 26
Cave(s), 42–43
Cellular concrete, 239–40
Cement. *See* Concrete
Chan, Wing-tai, 89
Chinese garden, 89
Chow, H. K., 244
Cinva ram, 151–52
Clark, Harold, 19
Clearstory: light in kitchen, 135; openings, 29, 335
Climate control, 23, 25–26; interior, 18. *See also* Climatology
Climatology: building, 18–24; and planting design
Closet(s), 130
Codes, building, 3
Collins, F. Thomas, 201, 204, 205
Color: and form, 339; harmony, 341; and light, 334–42; and psychology, 337, 338–40; reflection factors, 336–37; room, 339
Combination square, 262
Combustion, of firewood, 76, 77
Composite materials, 236–47; Bays's material, 243–44; bottles and cement, 246–47; burlap

368

sacks and cement, 241; cellular concrete, 240; clay, asphalt, and cardboard fiber, 244; concrete and bamboo, 244–45; corn cobs and concrete, 237; plastics, 248–53; rice hulls and cement, 237; sawdust-cement, 238

Composting, 320; aerobic, 320, 321; and compost privy, 320–21, 322–23, 327

Compost privy, 320–21, 322–23; sauna combination, 327

Concrete, 6, 192–200; air-entrained, 239; block, foundation, 273; block, spiral stairway-fireplace, 314 (See also Block[s]); and bottle construction, 246–47; decks, 308; ferro-cement, 194; floors, 280–84; forms, 196–98, 199, 202; and free-form house, 109; joists, 307; lightweight, 238–39; masonry, 182; mixing, 196; no-fines, 198–99; panels, precast (See Precast concrete panels); pier fittings, 272–73; plastered fiber, 240–41; reinforced, 192, 194, 241–46; roof, 305, 308–9; shell roof, 303; and shrinkage, 195; soil-cement house (Bogotá, Colombia), 8, 9; stairs, prefab, 312; starched, 241; strength, 195–98; and tile-block roofs, 302–5; wall forms, 196, 198, 199, 200; water content in, 195

Condensation zone, 293

Construction: building, 4, 5; earth, building ordinances against, 142; foundation, 276; pole-frame, 223, 231; rammed-earth, 158; tools, 261–66. See also specific types

Contraspatial house, 13–14

Convection, 19, 38

Conversation pit, 125, 126

Cooking: and dining, 131–35; and kitchen design, 132–35; range, heater, oven, wood-fired, 55

Cooling: cellar, 33; fans, 40, 41; night air, 22; north sky as, 21; roof, 37–40; summer, 25, 32–41; water, 32–33. See also Air conditioning

Corn cobs (ground), as concrete filler, 237

Corrugated shell structure, 242–43

Coelle, Jacques, 22, 246

Court-garden, 121; house, 106–15

Coverings, floor, 286–87

Crawl-space plenum, 71

Deering, Dr. Robert, 96

Design: and health, 128; home, 4; landscape, 88–97; and modular coordination, 120; rhythm, 121

Diatomaceous earth, 240

Dicker, Ed, 259

Dietz, Albert, 208–9

Dining: and cooking, 131–35; and kitchen space, 135

Direct lighting, 332–33

Do-it-yourself painting, 342–48. See also Painting

Drills, 261

Earth: block (See Block[s]); construction, building ordinances against, 142; diatomaceous, 240; floors, 280; mound construction, 110–11; nogging, 143; rammed (See Rammed earth); wall construction, 143; wall engineering, 143; wall test, 146. See also Soil

Earth-form lift slab, 111

Eckbo, Garret, 90

Edison, Thomas, 196–201

Entry passage, 124

Evaporation, 19

Expandable house, 104

Fabritz, Carl, 239

Fans, 40–41

Fiber concrete, plastered, 240–41

Fiber-reinforced plastics, 248

Financing, building, 2

Finsterlin, 106, 108

Fireplace(s): chimney design, 75–76, 78–81; and concrete-block spiral stairway, 314; corner, 83; draft, 76–77, 80; forms, 81–85; grate, 77; hearth, 77, 84; heat, 73–87; heat-circulating, 75, 81; location, 86–87; proportions, 82; stair-core, 315

Firewood combustion, 76, 77

Flagg, Ernest, 175, 176, 177, 240, 314

Floor(s), 277–87; concrete, 280–83; and rug dirt, 278; seamless polyester, 252; slab construction, 281–84; vents, 30; wood, 284–85

Fluorescent lighting, 332

Foam plastic: characteristics, 250; free-form earth-formed, 253; sprayed, 250; susceptibility to heat, 250; thermal insulation, 250

Folding armature dome, 252

Footing(s): concrete pier, 272–73; design, 269, 270; mat-type, 272

Forms. See specific forms

Foundation(s), 267–76; concrete-block, 273; construction practices, 276; footing design, 269; grade-beam, 273; layout, 275; stone-filled, trench, 273; systems, 272; wood, 274–75

Frame: pole, structures, 228–35; wood, structures, 219–27

370 · INDEX

Framing: balloon, 222; details, 223; systems, rational, 221. *See also* Wood
Free-form house, 106–15, 253; building, 109; and concrete, 109; and continuous shell construction, 109; curving, 109; plastering, 9, 109; and plastics, 250–51
Frost depth, maximum, 271
Fryer house, 177
Fuller, Buckminster, 225, 260
Furniture, 122; plastic, 252

Gaudi, Antonio, 106, 116–17
Gauger, Nicholas, 75
Geddes, Patrick, 11
Geiger, John, 196–97
Geiger, Rudolf, 92
Gill, Eric, 116
Glass. *See* Window(s)
Glen, A. L., 7
Goetheanum, 106
Goff, Bruce, 106
Greenhouse effect, 61
Group living: alcoves, 125; entry-passage planning, 124; plans, 123; space, 122–26

Hammers, 262–63
Harada, Dr. Jiro, 15–16
Heat: central, 63–72; and climate factors, 64; core, combination, 53; fireplace, 73–87; loss, 71–72; and physical comfort, 66; pipes in floors, 281; pump, 56–57; radiant, 67–68, 70; solar (*See* Solar); sources, 64; space, 52–62; transfer, and radiation, 38
Heaters: according to climate, 65; electric, 63; and fuel costs, 66; oil, 63; and plenum method, 70–71, 86; radiant, 67–68, 69, 70; wood, Ashley, 54; wood, owner-built, 55; wood, Riteway, 54–55
Heliodon(s), 92–95, 336; home-made, 94; use of, for plant locations, 94–96
Hollein, Hans, 106
Holsman, Henry, 190
House: court-garden, 116–21; design, and structural purpose, 116; free-form, 106–15; planning and building site, 13
House style(s), 98; free-form, 106–15, 253; and mobility, 99; and passive-active states, 101. *See also* Space
House wrecking, 255–56; tools, 256
Humidity, 19–20
Hurst, Homer, 230, 232–35, 285

Illumination. *See* Lighting
Incandescent lighting, 332

Individual-living space, 127–30; and privacy, 128
Inside your home, 136–40. *See also* Interior space
Insulation, 36–39; and floors, 283; foam-plastic, 250; masonry-block, 187, 188; materials, 294; stone-wall, 177; roof, 307–8; and vapor barriers, 292; and ventilation, 33, 36
Inter-American Housing and Planning Center (Bogotá, Colombia), 8
Interior space: and climate, 18; effect of, 136; and polyfunctional endo-space, 138
International exhibition on low-cost housing (New Delhi), 6, 7

Johnston, William, 60
Joistile roof, 305
Joists, 209, 225; precast, 307
Juhnke, Paul, 201, 204, 205

Kahn, Lloyd, x, 228
Kiesler, 105, 106, 108
Kira, Alexander, 320
Kirkham, John, 158; residence, 157–58
Kitchen(s): arrangement, 132–35; counters and cabinets, 134; and dining space, 135; light, 135; open, 135; storage, 132–35

Labor, building, 3
Ladell, Dr., 21
Laminated arches, 226
Lamps. *See* Light(ing)
Landscape design, 88–97; and Chinese garden, 89; and climatology, 92; and heliodon, 92–95; and microclimate, 92; procedure, 92; and Spieltrieb concept, 91; and variety planting, 89–90. *See also* Plant(ing)
Landscape and Living, 90
Laundry room(s), 130
Le Corbusier, 118, 119, 131
Levitt, 68
Lift-slab construction, 202–6
Light(ing): clearstory, 29, 135, 335; and color, 334–41; direct, 332–33; fluorescent, 332; and glare, 44; and health, 332; incandescent, 332; indirect, 331–32; mood, 333; psychological effects of, 45; spot, 333; types of, 137, 331; and wiring, 328–33. *See also* Wiring
Light and shade, 42–51; effects of, to scale, 121
Lightweight aggregates for building, 239–40
Lindstrom, Rikard, 322
Lithosphere building, 22
Living space: group, 122–26; individual, 127–30. *See also* Space

Loans, building, 2
Loft(s), 129
Log: building, 226, 227; cutting and seasoning, 215. *See also* Wood
Loomis, Mildred, ix
Louvers, as vents, 30
Lumber. *See* Wood

Machine-for-living approach, 14, 15
Magdiel brothers, 162, 177, 199
Maillart, Robert, 193
Marx, Roberto Burle, 91
Masonry; block, 179–91 (*See also* Block masonry); brick 179–91 (*See also* Brick); roofs, 301–9 (*See also* Roof[s]); stone, 165–78 (*See also* Stone)
Materials, 3; comparison of properties, 251; composite, 236–47; salvage, 254–60. *See also specific materials*
Maybeck, Bernard, 241
Mendelsohn, Eric, 106
Microclimate, and landscape design, 92
Middleton, G. F., 7
Minimal-Cost Housing Group, 8, 9
Moisture barrier: and floors, 281; polyethylene, 253
Moller, Clifford, 254, 255
Mortar. *See specific types*
Mother Earth News, The, ix

Nail(s), 208, 209, 211, 212; and holding power of wood, 210; popping, 211
Natural: lighting, 43, 45; resources, 4; ventilation, 29
Nearing, Helen and Scott, 176, 177
Nervi, Pier, 193, 194
Neutra, Richard, 17, 267
Niemeyer, Oscar, 106
No Fango method of strengthening concrete, 241

O'Gorman, Juan, 106
Olgyay, Aladar and Victor, 93
Open plan, 102–4; and furniture, 122; and group-living spaces, 122–24
Ortega, Alvaro, 201, 202
Osmond, Dr. Humphrey, 119, 120
Ott, John, 42, 332
Owner-builder, seven axioms, 2–5
Owner-Builder Publications, x
Owner Built Homestead, 11

Paint: alligatoring, 346; blistering, 346; brushes, 348; cement-water, 347; exterior, 345–47; formulas, 345–46; latex, 346–47; synthetic, 344
Painting: decorative, 343; do-it-yourself, 342–48; surfaces, 348
Panel construction, 224–25
Panels: Plastic wall, 252; precast-concrete, 201–7; stressed-skin, 224–25, 300. *See also specific panels*
Patch, O. G., 198
Patterson, F. C., 187
Patty, Ralph, 162
Peters, Frazier, 176, 177
Pichler, Walter, 106
Pipe(s), plastic, 250, 252, 253
Plan, 98–105; building flexibility, 103; and concepts of space, 100–101; and concept of freedom 100–101
Planes, 264
Plank-and-beam structural system, 222, 298–99
Plant(ing), 89–90, 91–92; design procedure, 92; effect on surrounding atmosphere, 96–97; selection of, 96–97; for weather protection, 92. *See also* Landscape design
Plaster: cement, 109; and lath, nonbearing partition walls, 240; proportioning, mixing, and application, 110; tools, 110
Plastic(s), 248–53; bath units, 252; celluloid, 250; foam, 250–52; piping, 250, 319; polystyrene, 250; polyurethane, 250; sealants, 252; sprayed urethane foam structure, 249; thermoplastics, 250; thermosetting resins, 250
Pleijel, Gunnar, 21
Plenum, 70–71, 86
Plumbing, 316–27; codes, 316–17; compact, 318; costs, 319; fixture arrangement, 318; minimum, 318; plastic piping, 250–319; stack, 317–18
Plunger-pile floor, 283
Plywood: floors, 286; gusset plate, 297; structural skin panels, 224–25, 300. *See also specific uses*
Pole-frame: and adaptability, 231; continuous-structure building, 234–35; details, 233; embedding, 231–32; house, 229; poles, 258; and site conditions, 230; structures, 228–35; and treatment of poles, 232
Polyethylene, 252–53
Polystyrene, 250
Polyurethane, 250
Post and beam. *See* Plank-and-beam structural system
Post-and-girder structural system, 222

Power tools 263–64, 265
Precast concrete panels, 201–7; rigid-frame, 206; tilt-up, 202–7; vacuum-process method of making, 201–2
Prefabricated: concrete panels, 201–7; concrete stairs, 312; framing systems, 223; stressed-cover panels, 224–25; wall units, wood, 223–24
Pressed block, 150–56; box test, 153–54; curing, 153; mixing proportions, 153; presses, 150–52; structural tests, 154

Raabe, John, vi
Radiation, 19; effective, 78; and heat transfer, 38
Rafter(s), 225; systems, trussed, 297–98
Rammed earth, 6, 157–64; compression strength, 158; construction, 158; forms, 159–61, 162; stabilizers, 162; wall building, 159; wall finishes, 162–63; waterproofing, 162
Reinforced: block, 189–91; concrete, 192, 194, 206–7; concrete, with bamboo, 244–45; concrete, fiber-reinforced, 240–41; concrete, plastics, fiberglass, 248; concrete, with waste iron and tin, 244; concrete, with wood, 245–46
Resources, natural, 4
Ricci, Leonardo, 106
Rice, John, 239
Rice hulls and cement block, 237–38
Roberts, Rex, 39
Rock. See Stone
Roof(s): block, 305; butterfly, 29; and cooling, 37–40; evolution, 296–97; gardens, 119; Joistile, 305; masonry, 301–9; plank-and-beam, 222, 298–99; and roofing temperatures, 39; sag, 299; sod, 309; steep, 308–9; tile-block and concrete, 302–5; truss-type, 297; water-cooled, 39–40; wood, 296–300; Zed-tile, 306
Rosin, Dr. P. O., 79–80
Rub-R-Slate, 243
Rudofsky, Bernard, 91, 278–79; walls, 90
Rug and floor dirt, 278
Rumble, Roy, 200
Rumford, Count, 75–76, 78, 79

Salvage materials, 254–60; burlap sacks, 241, 258; car bodies, 259; ground corn cobs, as concrete filler, 237; railroad ties, 258; rice hulls and cement, 236; sawdust, 239; sulfur block, 8
Sampson, Foster, 335

Sauna, 323–24; compost privy combination, 327
Savot, Louis, 75, 77
Sawdust-cement composite, 238–39; floors, 280
Saws, 263
Scale: in building design, 120; effect of light, shadow, and spatial sequence to, 12
Schmidt, Carl, 166
Schmidt, Dr. Ernst, 22
School of Living, ix, x
Scott, Geoffrey, 121
Screens, 28
Screwdrivers, 264
Sealants. See specific materials
Septic: disposal fields, 325; systems, 324, 326; tanks, 320, 324; tanks, design, 325
Service area, 17
Sewers, 253, 319
Shade and light, 42–51
Shading: devices, 50–51; efficiency of methods 49; external, 48
Shell house, 242; roof, concrete, 303
Shutters, 46
Shuttleworth, John, ix
Siple, Dr. Paul, 35
Site. See Building site
Skylights, 335
Slab. See specific slabs
Sleeping lofts, 129
Snow loads, maximum, 299
Sod roofs, 309
Soil: load-carrying capacities of common, 267–70; tests, 145–46, 270. See also Earth
Soil-cement house (Bogotá, Colombia), 8, 9
Solar: collectors, 59–60; exposure and court-garden house, 118; heat exclusion, 50–51; heating, 22, 57–59; heating, sun-tempered, 60–61; heating, supplemental, 60; orientation and air conditioning, 37; wall, 61–62; water heaters, 326–27
Soleri, Paolo, 106; desert home, 110–11
South African Research Institute, 8
Space: endo-, ecto-, and meso-, 139; flowing, open planning, 102–4; freedom, 100–101; group-living, 122–26; heat, 52–61; individual, 127–30; and motion studies in homes, 128; operative, 101; private, 101; public, 101; qualities, 100–101; sociofugal, 120; sociopetal, 120; time relationship in design, 118
Spieltrieb concept in garden plan, 91
Spotlights, 333
Sprayed urethane foam structure, 249
Squat-type toilets, 320
Stack effect, 27

Stack sack, 259
Stains, 344, 345, 346
Stair(s), 310–15; prefab concrete, 312; rules for building, 310; spiral, concrete-block, fireplace, 314; spiral, low-cost, 312–13; straight-flight, 311; wood, 311
Starched concrete, 241
Stein, Joseph Allen, 7
Steiner, Rudolf, 106
Stone: bedding, 167; choosing, 168–70; classification, 168; construction properties, 169; forms, 175–76; grain, 167; laying, 171–75; locating for use, 166; masonry, 165–78; mortar for laying, 170; rift, 167–68 seams, 167; strength, 170; tools; 171; wall, 173; wall, insulation, 177; work, 172; workability, 170–71
Storage: closets, 130; kitchen, 132–35
Stove(s), Franklin, 53. See also Heaters
Straw in adobe blocks, 144
Stressed-cover panel, 224–25, 300
Stud-wall construction, 222
Style, 3
Subterranean-tempered air conditioning, 22
Sulfur: blocks, 8, 202; floor tiles, 9, 10
Summer cooling, 25, 32–41; night air, 22; north sky as, 21; and summer breezes, 31. See also Cooling; Air conditioning
Sun. See Solar

Tabor, Dr., 62
Tamped earth, 143–44
Temperature, 19–20, 21, 26; and convection, 19; corrected effective (C.E.T.), 20; effective, 20; and evaporation, 19; and humidity, 19–20; and radiation, 19
Thermoplastics, 250
Thermosetting resins, 251
Thomas, Wendell, ix, 22, 55–58, 70
Tile: block and concrete roof, 302–5; Zed, roof, 306–7
Tilt-up, 202–7; Collins, 204; erection in wood framing, 223; Juhnke, 205
Toilet(s), 318, 319, 320; and health, 320; squat-type, 320
Tool(s), 4, 261–66; accessories, 266; cement-slab, 284; house-wrecking, 256; masonry, 171; plaster, 110; storage, 265
Trifunctional house, 13–14
Trombe-Michel solar wall, 61–62

Van der Rohe, Mies, 117
Van Dresser, Peter, 22, 60
Van Eyck, Aldo, 136, 139

Van Guilder, 196
Vapor barriers and location of insulation, 292
Venetian blinds, 335
Ventilation, 25–31; and attic, 30, 40–41; and insulation, 33, 36; natural, 29; openings, 28; screening to reduce, 28; stack effect, 27; wind scoops as, 28
Vents: continuous-ridge, 31; in external walls, 28; floor and attic, 30; louvers as, 30; and plenum design, 70–71, 86; rake, 31; roof, 31; sewer, 317; soffit, 31
Venturi, Robert, 137
Vermiculite, 240

Walker, Major de W., 240
Wall(s), 288–95; block, 183; bracing tests, 290; building, rammed-earth, 159; circular, 113; concrete, 198–200; curved, building forms, 114; durability, 291; finishes, rammed-earth, 162–63; finish and lighting, 332; forms, concrete, 196–99; forms, rammed-earth, 159–61, 162; forms, stone, 175–76; foundation, 270; function, 288–89; masonry (See Block masonry); material selection, 289–90; moisture, 291; planning, 292–93; pole-frame, 229–30; spiral, slip-form, 115; stone, 173; stud, construction, 222; thermal performance, 293; tilt-up, 202–7; units, prefabricated, 224; vents, 28
Warm-air radiant-convector, 70
Waste disposal, human. See Compost privy; Septic
Water-cooled roofs, 39–40
Water heaters, solar, 326–27
Watson, George, 244
Whole Earth Catalogue, The, x
Wilson house (Coursegold, CA), 315
Wind, 27–28; breaks, planting as, 92; breaks and heating efficiency, 64; scoops, 28
Window(s) 15, 28; and air movement, 28; clearstory, 29, 135; double- and triple-glass, 46–47; frames, plastic, 252; ill-chosen, 28; and lighting, 44–46; openings, 27; picture, 15; protection, 21; sashes, 47; sill, block, 184; thermopane, 46; unit, low-cost, 48
Wiring: and lighting, 328–33; methods, house, 329; procedures, 329–30. See also Light(ing)
Wood, 208–18; bending strength, 216–17; characteristics, 209–10; classification, 215–18; decay, 216; fasteners, 212–14; floors, 284–85; foundations, 274–75; frame structures, 219–27; frame walls, 291; framing details, 223; framing systems, 221; framing systems, prefabrication, 223–24; framing

systems, rigid, 223; gluing, 214; hardness, 215–16; laminated arches, 226; log building, 226–27; log cutting and seasoning, 215; louvers, as vents, 30; moisture content, 210–11; pole-frame structures, 228–35; as reinforcement in concrete, 245–46; roofs, 296–300; salvaged, 25; shrinkage, 210–11, 216; siding, exterior, maintenance, 344–45; stairs, 311; stiffness, 217; strength, 214;

swelling, 216; toughness, 217; warping, 216; wear resistance, 217
Workshop, home, 265–66
Wright, Frank Lloyd, 25, 43, 67, 103–5, 111, 124, 136, 140, 188, 230, 255, 273, 276, 337

Zed-tile roof, 306–7